Hidden In Plain Sight 2

Andrew Thomas studied physics in the James Clerk Maxwell Building in Edinburgh University, and received his doctorate from Swansea University in 1992.

He is the author of the *What Is Reality?* website (www.whatisreality.co.uk), one of the most popular websites dealing with questions of the fundamentals of physics. His first book, *Hidden In Plain Sight*, is a science best-seller.

Also by Andrew Thomas:

Hidden In Plain Sight
*The simple link between relativity
and quantum mechanics*

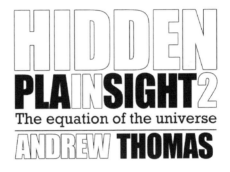

Hidden In Plain Sight 2: The Equation of the Universe

Copyright © 2013 Andrew D.H. Thomas

All rights reserved.

ISBN-13: 978-1479294411
ISBN-10: 1479294411

DEDICATION

To Mum and Dad

CONTENTS

1 Introduction   1
    Simple solutions to big problems
    Forces
    The principle of least action
    Fundamental principles
    Structure of chapters

2 Gravity   17
    A history of gravity
    General relativity

3 Cosmology   35
    The scale of the universe
    Olbers' paradox
    The Big Bang
    The shape of the universe
    Inflation

4 Dark Matter and Dark Energy   59
    Dark matter
    MOND
    Dark energy
    The accelerating expansion
    The biggest error in science

5	**Black Holes**	79

Supermassive black holes
Black hole thermodynamics
Information and entropy
The holographic bound
The information loss paradox

6	**The Soap Bubble Universe**	103

The universe must be flat
The hybrid expansion

7	**Gravity Revisited**	121

The accelerating expansion (revisited)

8	**How to create your own Universe**	133

Is the universe a black hole?
Enter the black sphere

9	**Conclusion**	145

PREFACE

This is my second book. I was absolutely delighted at the success of my previous book, *Hidden In Plain Sight*. If you bought my previous book, then thank you very much. If you have not already read *Hidden In Plain Sight* I can recommend it as some of the work in this book builds on the principles described within it.

The discussion and reasoning in this new book might appear to be extraordinarily simple. This is quite deliberate. As I explained in my previous book, I am convinced that the fundamental principles which underlie reality **must** be simple. The drive to uncover the fundamental truths of physics must surely be a drive toward simplicity.

After considering the link between quantum mechanics and relativity in my previous book, this new book considers a possibly simple solution to the unsolved mysteries behind gravity, dark energy, black holes, and many other cutting-edge problems in current theoretical physics. My books are nothing if not ambitious!

I deliberately make my books fast-paced and easy to read, packed with only relevant information and no padding.

So it's a book with a real big bang!

Andrew Thomas (hiddeninplainsightbook@gmail.com)
Swansea, UK,
2013

grav·i·ty (grăv′ĭ-tē)
n.
> 1. The natural force of attraction between any two massive bodies.
>
> 2. A serious situation or problem.

1
INTRODUCTION

As physics progresses along its path of unlocking the secrets of the universe, it might be thought that the number of unsolved problems would be reduced. However, this does not seem to be what is happening. Instead, the more we find out about the universe, the more problems we unearth. And the problems we discover seem deeper, more profound, and more insoluble than the problems we have just solved. This was summed-up by the great physicist John Wheeler: "We live on an island surrounded by a sea of ignorance. As our island of knowledge grows, so does the shore of our ignorance."

It has been suggested that the field of theoretical physics seems to be in less-than-superb shape at the moment. The main thrust of research towards a possible "theory of everything" is directed towards string theory. This may well eventually emerge to be the correct model of Nature, and a great deal of very clever people seem convinced by it, however, considering the amount of research effort which has been directed at this theory over more than four decades, it is hard to disagree that progress has been disappointing. But there appears to be a lack of diversity of alternative research.[1] String theory has been called "the only game in town", and maybe that is the heart of the problem.

Meanwhile, the highly-publicised discovery of the Higgs boson by the Large Hadron Collider (LHC) seems to have created panic among particle physicists. You might find that surprising, after all, didn't they predict that the Higgs boson was there to be found? Well, yes, they did – and that's the problem. An experiment which just verifies your prediction tells you nothing new about reality. It so happens that the properties of the discovered Higgs boson were a particularly close match to those predicted. Eight billion euros to tell you something you already knew does not sound like good value for money. There is no hint in the LHC results of new directions for research, and hence there is no convincing motive to spend new money. This sounds very much like the end of the line. A lot of particle physicists are scared for their jobs.

It appears that the LHC has also brought bad news for string theory. The full name of string theory is "superstring" theory, with the "super" standing for "supersymmetric". According to supersymmetry (or SUSY), each elementary particle is paired with a superpartner particle. However, the LHC has not detected any of these predicted particles.

If I had to sum up the current state of theoretical physics, I would say it lacks direction. It appears that we have come to the end of a long road – and we have no map to tell us which direction to take next.

---

[1] Things seem to be changing, with only 9% of U.S. faculty hired in 2012 being string theorists, compared with 58% in 1999.

INTRODUCTION

So where do we go from here? Well, if you are like me, you got interested in physics because you want answers. You want to get to the bottom of things. We want to know "why" things happen. However, most physics research is merely aimed at **describing** the phenomena of Nature – it is not aimed at answering fundamental questions of "why". Such questions would be considered the domain of philosophy, not physics.

A good example are the equations of general relativity, which describe gravity as emerging from the curvature of spacetime. These equations are clearly an improvement in accuracy over Newtonian gravity, but seem to get us no further along the road of explaining just **why** masses attract each other. Saying that masses follow straight lines in a curved spacetime does not answer the fundamental question of **why** objects follow a straight line in curved spacetime. We are just describing the phenomena – not explaining it.

And so many of these descriptions seem excessively complicated to my eyes, string theory in particular seeming to revel in excessive mathematical virtuosity. It seems as though we are moving away from a search for simplicity in favour of increasing complexity. Is this really the likely route to unlocking the secrets of the universe? I feel very strongly that as we dig deeper, and approach more fundamental levels, then our theories should actually become simpler and easier to understand. My books reflect this belief.

The majority of current research seems only interested in capturing the minutiae of the complexity of Nature – there is no consideration of the bigger picture. I would agree with Sir Patrick Moore, the sorely-missed presenter of the long-running BBC television programme *The Sky At Night*:

*The one inescapable fact is that we exist, and so does the Sun, the stars, and the Earth, and everything else. And no one has yet explained how the matter came into existence in the first place. Which adds force, I think, to my own contention that **we are pretty strong on detail but we are still very weak on fundamentals.***

In my series of books, I am trying to get to the bottom of things. By working with fundamental principles it appears we can get satisfactory answers to questions of **why** Nature behaves the way it does.

Simple solutions to big problems

We could list several of the most puzzling and important unsolved problems in current theoretical physics:

- The unification of quantum mechanics and relativity. The two dominant theories of physics remain stubbornly at odds with each other. However, it should be possible to combine them into a single, deeper theory.

- How can we make sense of the "weirdness" of quantum mechanics? What does it all mean?

- Dark matter. What is the mysterious substance which greatly exceeds the amount of normal mass in the universe?

- Dark energy. What unknown force is powering the acceleration of the expansion of the universe?

INTRODUCTION

- The black hole information loss paradox. General relativity says that information can be lost in black holes, but, according to quantum mechanics, this should be impossible.

- Inflation of the universe. The theory which has been at the cornerstone of cosmology for thirty years has been found to be flawed. But any replacement theory has to be able to explain why the structure of the universe is so smooth.

The first two problems I have listed – the unification of quantum mechanics with relativity, and how to make sense of bizarre quantum mechanical behaviour – were considered in my previous book. I am pleased to say that this new book is going to propose a potentially simple solution to many of the remaining problems.

Forces

Why do things move?

That's a fairly profound question. Why do apples fall? Why do the planets orbit the Sun? Why do magnets attract pieces of metal?[2]

The general answer to the question of why things move would be to say that they have been acted-on by a force. It was Isaac Newton who first correctly identified the role of a force in producing motion. If a force is applied to a stationary object, then the object will start to move. The force acts to accelerate the object.

There are four fundamental forces: the electromagnetic force, the strong nuclear force, the weak nuclear force, and gravity. These are all *conservative* forces. A conservative force does not lose energy as it acts (hence, "conservative"). This means that if an object is moved in a closed loop by a conservative force, and thus returns to its starting position, the net work done is zero.

As an example of this feature of a conservative force, imagine throwing a stone directly upwards at a speed of ten miles per hour. The stone will gradually lose speed as it rises, and will come to a halt at a maximum height. The stone will gain speed as it descends and, when it passes your hand on the way down, it will again be moving at ten miles per hour.

---

[2] You may be aware of the work of the young American gentlemen known as the Insane Clown Posse, a so-called "hip-hop" group who wondered about the nature of the electromagnetic force in their track "Miracles" (video available on YouTube).

INTRODUCTION

This conservation of the speed of the stone is due to gravity being a conservative force.

However, this will only be the case if we ignore the effect of air resistance, which is a form of friction. Friction is an example of a non-conservative force. Friction acts to slow the speed of the stone, to decrease the energy of the stone, and to dissipate that energy into the environment in the form of heat. For example, if you throw a stone up a sand dune at ten miles per hour, the stone will find the sand sticky due to friction, and when the stone passes your hand again on the descent, its speed will be considerably less than ten miles per hour. The energy of the stone will have been dissipated into the environment.

Conservative forces (i.e., forces without friction) are perfectly reversible. For example, if you made a video of snooker balls knocking into each other on a snooker table, and you reversed that video, it would still look like normal behaviour. Even when the action is reversed, you would still see balls hitting other balls and rebounding according to Newtonian mechanics. However, non-conservative forces are not reversible. This is because non-conservative forces dissipate energy into the environment. Eventually, the snooker balls would slow to a halt due to the effect of friction. This is an irreversible effect.

The real reason behind the irreversible nature of non-conservative forces is due to a quantity called *entropy*. Entropy can be thought of as the amount of disorder or randomness in a system. We will be considering entropy in detail in Chapter Five, but for now all we need to know is that the entropy of a system will always increase, and can never decrease. This seems to make a lot of sense: a system tends to increasing disorder over time. For example, your glistening new car will wear-out over time as its atoms become increasingly disordered through wear and tear. The non-conservative force of friction converts energy into heat, which is the random motion of atoms. This randomness of heat means it has a high entropy. This is what introduces the irreversibility of friction: the second law of thermodynamics states that this increased randomness and disorder of the heat cannot be reversed. So just as your old car can never reverse its wear and tear to become shiny and new again, so a non-conservative force such as friction cannot be reversed.

The principle of least action

There is clearly a connection between forces and energy. This is because a moving object possesses *kinetic energy*. So a force which produces motion in an object actually transfers energy to that object.

As an example of the connection between energy and motion, consider dropping a ball from your hand. When the ball is in your hand, it stores *potential energy* within itself. Potential energy is the energy an object possesses purely due to its location. When the ball is held in your hand, its location is a certain height above the surface of the Earth, so the ball contains gravitational potential energy. This is not the only type of possible potential energy. For example, a stretched spring can contain elastic potential energy.

When you let go of the ball in your hand, it picks up speed. So the kinetic energy of the ball is increasing. However, it is falling closer to the surface of the Earth so its potential energy is decreasing. So the potential energy of the ball is being converted to kinetic energy as the ball accelerates ever faster.

There seems to be a balance here, in that the amount of potential energy in the ball is being reduced by precisely the same amount that its kinetic energy is increasing. This is what we would expect due to the law of conservation of energy which says that the total amount of energy in a closed system remains constant over time.

INTRODUCTION

We could ask the question, why is there this conversion from potential energy to kinetic energy? After all, the total energy of the system is going to be unchanged. So, if that is the case, why does Nature bother doing the conversion from potential to kinetic energy? What is the point?

Well, that's a good question. And, as we shall see in this book, I think the answer holds an important key to understanding the universe. All we can really say at the moment is that Nature tries to reduce potential energy to its minimum possible value. This has been called the *principle of minimum potential energy*. Nature appears to have something of an obsession with reducing potential energy.

As an example of this principle, consider a soap bubble. A soap bubble is an extremely thin film of soapy water which contains a volume of air. That thin surface has elastic properties, so, like the spring example, we could consider the surface as having elastic potential energy. It turns out that the potential energy contained in the soap film is proportional to its surface area. Nature tries to reduce this potential energy to a minimum, so Nature tries to reduce the surface area to a minimum – bearing in mind that the bubble still has to enclose a certain volume of air. As a result, soap bubbles are spherical, a shape which represents the minimum possible surface area for a given volume.

So the shape of soap bubbles is determined by Nature trying to reduce potential energy. The resultant shape is an elegant, efficient curve – we do not find soap bubbles with wasteful, ostentatious corners and spikes. Nature abhors waste and complexity and will always prefer a smooth, simple, elegant solution – purely because Nature always tries to reduce potential energy.

In the field of mechanics (the science of moving bodies), it is known that this minimisation of potential energy, and its conversion to kinetic energy, is the key to analysing the movement of a system – no matter how complicated that system might be. If we consider a moving system, possibly composed of many parts, we can calculate the difference between the kinetic energy of the system minus its potential energy. This value is known as the *Lagrangian*.

We can calculate the Lagrangian of a moving system at any point in time to get a series of values. If we then sum all those values we get what is known as the *action*. The action is a very important concept which is often used by physicists, but is not very well known by non-physicists. The action is important for our purposes because it appears to tell us something very deep about the behaviour of Nature.

If we consider the action of our moving system over a period of time, we will find that the value of that action will always be the lowest possible value. It is as if Nature always moves objects so as to minimise the action. This very important and useful principle is called the *principle of least action*.

As an example, consider the following image of the trajectory of a ball which has been thrown:

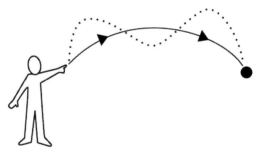

I am sure you can tell that the ball will follow the smooth curve of the black line – it would never follow the irregular dotted line. Just as with the soap bubble example, Nature prefers smooth curves to wasteful twists and turns.

We can calculate the Lagrangian of the motion of the ball (remember: the Lagrangian is the difference between the kinetic energy and the potential energy) at each point, and then we can sum all of those points to calculate the action.

INTRODUCTION

The principle of least action tells us that the action will be the smallest possible value for the path taken by the ball. In this case, the smooth path taken by the ball results in the smallest action.

As another example, imagine you are a long-distance runner trying to decide on a strategy for your next race. You could start off very quickly, running the first half of the race very fast, and just trying to hang-on as you slow down in the second half of the race. Or else you could start more conservatively, running the first half of the race slower but putting in a sprint to run the second half faster. But, as any distance runner will tell you, the strategy which will result in the quickest finishing time is an even-paced race, running both halves of the race in the same time. So, once again, we find that the smooth trajectory with minimal variations is the most efficient strategy and wastes least energy. Just as the long-distance runner tries to minimise his energy output and so selects a smooth trajectory, so does Nature try to conserve energy and also selects the smoothest trajectories. This is analogous to the principle of least action.

We can use the principle of least action to predict how a system will behave. We know that the system will always move in such a way that the smallest action will result. This is a very useful way to analyse and predict the behaviour of a mechanical system. The Lagrangian approach is a very flexible approach which allows a variety of widely-differing systems to be analysed.

However, action remains something of a mystery. This was summed-up by Robert Matthews talking about action in a *New Scientist* article:

> *For reasons as yet utterly mysterious, this quantity stays as small as possible under all circumstances. Theorists are convinced that action must be incredibly important – so much so that the discovery of any new fundamental law prompts a race to work out the*

particular action needed to produce it. The trouble is that no one understands the principles behind Nature's infatuation with action. [3]

Might it be possible to uncover this principle behind "Nature's infatuation with action"? There seems to be some principle underlying Nature's obsession with the efficient reduction of potential energy. This principle seems to lie at the heart of motion in the universe, and therefore the evolving structure of the universe itself. As Robert Adair says in his book *The Great Design: Particles, Fields, and Creation*:

If Nature has defined the mechanics problem of the thrown ball in so elegant a fashion, might She have defined other problems similarly? So it seems now. Indeed, at the present time it appears that we can describe all the fundamental forces in terms of a Lagrangian. The search for Nature's One Equation, which rules all of the universe, has been largely a search for an adequate Lagrangian.

The principle of least action makes it clear that it is the balance (and imbalance) of energies which is the key factor in determining the motion of objects, and it is this balance of energies which we will be considering in the later chapters of this book.

---

[3] Robert Matthews, *I is the Law*, New Scientist, 30th January 1999.

INTRODUCTION

Fundamental principles

Let's return to the fundamental question of why things move. In the 1973 film *The Rat's Death*, a woman slips over her child's toy, so she slaps the boy, the boy kicks the dog, the dog barks at the cat, so the cat pounces on a rat. So if we try to find an answer to the question "Why did the cat move?" we very quickly find ourselves moving progressively down a chain of cause and effect. Movement inspires other movement. We might then start to wonder what could possibly come at the end of the chain. What could possibly start this chain of causality, or does it simply progress forever, an infinite regression?

A similar question was asked in my previous book, *Hidden In Plain Sight*. In that book, we considered the unification of two different theories to create a simpler theory. For example, Michael Faraday unified electricity with magnetism to produce electromagnetism. The process of unifying theories was compared to moving down a tree, with twigs merging into branches, and the branches merging into the trunk. As we move down the tree, we find our theories getting fewer and simpler. The question then arose, what could possibly lie at the base of the tree? What could possibly hold up the entire structure? We would surely expect to find a principle which is simple and elegant. As John Wheeler said: "To my mind there must be, at the bottom of it all, not an equation, but an utterly simple idea. And to me that idea, when we finally discover it, will be so compelling, so inevitable, that we will say to one another: 'Oh, how beautiful. How could it have been otherwise?'"

The fundamental principle at the root of John Wheeler's tree would have to be strong in the sense that it could never be refuted – it had to be obviously correct. It would have to contain within itself the reason for its obvious correctness. It would have to be impossible to conceive of any universe in which it could not be true. For this reason, a fundamental principle could form the bedrock of any fundamental theory of the universe.

In my previous book, I decided that the principle that there was "nothing outside the universe" could be considered as being a fundamental principle. This is obviously correct: if the universe is defined as being everything that exists, then it could not be possible for anything to exist outside the universe – by definition.

Likewise, as we move down our chain of causality of moving objects, we seem to require some fundamental principle to lie at the base. From our earlier discussion of potential and kinetic energy it would appear that change of energy is intimately associated with the question of why things move, so we would probably expect our fundamental principle to have something to say about the balance of energies.

We will returning to consider this point later on in the book.

INTRODUCTION

Structure of chapters

The force of gravity plays a central role in this book. A fuller understanding of gravity will be seen to be the key to many problems in physics which appear mystifying at the moment.

Chapter Two provides coverage of our current knowledge of gravity, together with a history of how the main principles were discovered.

The importance of gravity in shaping the structure of the universe is a key aspect of this book, so Chapter Three will introduce cosmology, the study of the universe as a whole. We will be considering the main principles of the Big Bang theory, the shape of the universe, and the inflation hypothesis.

Chapter Four describes the mysterious dark matter and dark energy which seem to pervade the universe, dominating ordinary matter. Dark energy in particular is going to play a central role in the book.

Chapter Five will consider another mystery of the universe: black holes. The more we learn about black holes, the more important they appear to be. We will find out how they hold galaxies together. We will learn about the peculiar role of information and thermodynamics in black holes. We will discover how black holes seem to behave like holograms, and how the loss of information in black holes represents a tremendous paradox for physics.

In Chapters Six, Seven, and Eight we start to find potential solutions to all these mysteries. It appears that the problems of dark energy, the black hole information loss paradox, and many other problems associated with inflation could potentially be solved by a remarkable innovation.

2

GRAVITY

This is a book about gravity.

We are all well-acquainted with the force of gravity. Of all the four fundamental forces we only have to deal directly with gravity in almost all our daily activities. From an early age, we are taught that "things fall down". In this respect, we tend not to even think of gravity as a force – it is just a fact of life.

Gravity is, in many ways, an unusual force. Of the four fundamental forces, gravity is the only force which is always attractive for all objects. This is not the case with, for example, the electric force which treats negatively-charged and positively-charged objects differently (repels or attracts accordingly).

Another factor which separates gravity from the other fundamental forces is its extraordinary weakness. Out of the four fundamental forces, gravity is by far the weakest force. The force of gravity is actually a thousand million billion billion billion times weaker than the electromagnetic force. The Earth is a huge mass of six thousand billion billion tonnes, yet it is very easy to resist its gravitational pull on an object by, say, lifting an apple from its surface. All that mass, but the force is so weak. The reason behind the weakness of gravity is considered one of the great mysteries of modern physics.

However, even though it is the weakest force, gravity is the dominant force in the universe. It is gravity which holds the planets in orbit. It is gravity which forms the stars and galaxies and black holes. It is gravity which powers nuclear fusion in the Sun to produce our daylight. It is gravity which shapes the universe. So why should the weakest force turn out to be the dominant force in the universe?

Ironically, it turns out that gravity is the dominant force precisely because the other forces are so strong. The other forces are tied-up keeping particles bound together inside the atom. We do not realise the extraordinary forces which exist within the nucleus of each atom. We only truly realise the energy possessed by the strong nuclear force during a nuclear explosion. There is extraordinary energy inside an atom, but it is kept on a leash by the strongest of forces.

Considering the electromagnetic force, for example, the positive and negative electric charges are tightly-bound within atoms (electrons attracted to protons). Hence, the positive and negative forces are usually perfectly balanced. The net result is that the overall electric charge of most atoms is neutral. So, to an observer outside the atom, there is no overall electric charge detectable outside the atom — despite the great strength of the forces inside the atom. And there is no overall net electric charge of the universe.

Gravity is different. Gravity treats all objects in the same way, and is an attractive force for all masses. There is no concept of "positive" and "negative" charge for gravity. There is no such thing as "negative" mass — no objects fall upwards. Hence, even though gravity is a weak force, it is also a cumulative force, and when you consider how much mass there is in a star, for example, it is easy to see why gravity becomes so important.

Our best model of gravity is general relativity. This is a beautifully elegant theory which can be used to predict the trajectories of objects under gravity with great accuracy. The theory is one of the greatest achievements of the human intellect. The theory says that gravity is a by-product of the curvature of space caused by the presence of mass. But the theory has nothing to say about why this curvature occurs.

Out of the four fundamental forces, only gravity is not included in our best "theory of everything" – the standard model of particle physics. It is the weakest, dominant, always-attractive force, but it is also the most puzzling.

We really don't understand gravity.

A history of gravity

What is physics? Physics comes from the ancient Greek word *physika*, which means the science of natural things, and it is there in ancient Greece that our story begins.

In the fourth century BC, the Greek philosopher Aristotle asked the question: "Why do objects move toward the Earth?" His answer was that "Objects yearn to be united with the Earth". In other words, Aristotle did not imagine the existence of an attractive force, instead he imagined there was some intrinsic feature of an object which compelled it to move downward toward the centre of the universe, which was its natural place. The Greeks gave a name to this tendency: "gravity". However, Aristotle believed the sky was the natural place for light objects, which explained why light objects floated upward. The Greeks also had a name for this light-hearted tendency: "levity".

Aristotle's ideas held sway for almost 2,000 years until Galileo challenged these notions. Copernicus had just shown that it was the Earth which revolved around the Sun – not vice versa. This indicated that the Earth was not the centre of the universe, as suggested by Aristotle. But when you drop an object, it does not fly to the Sun. This indicated that Aristotle's idea about objects always seeking to fall to the centre of the universe was wrong.

Aristotle also argued that heavy objects fell faster than light objects. Again, Galileo challenged Aristotle's ideas with a thought experiment. Imagine two balls dropped from a great height, one of the balls being significantly heavier than the other ball. According to Aristotle, the heavier ball would fall faster than the light ball. Galileo then wondered what would happen if the two balls were tied together by a rope. The obvious answer would be that the slow-falling ball would hold back the faster ball, and the resultant speed would be an average of the previous two velocities. However, according to Aristotle, the two balls joined together would represent a heavier object which would therefore fall faster than either of the balls previously fell individually. This appeared to be a logical flaw in Aristotle's argument. Galileo realised that the only way out of this problem would be if the two balls fell at identical speeds, i.e., the weight of an object does not determine how fast it falls.

Galileo is often called the "father of modern science" because he tested this hypothesis by experiments (supposedly dropping objects off the Leaning Tower of Pisa). This showed that, indeed, light objects and heavy objects both accelerated at the same rate when they fell, any difference in the rates of descent being due to air resistance.

This principle was dramatically demonstrated in 1971 when Apollo 15 mission commander David Scott dropped a geological hammer and a feather on the surface of the Moon. The lack of atmosphere meant a complete lack of air resistance, and both objects clearly fell at the same rate. (YouTube: "The Hammer and the Feather". Ignore the comment that the feather was made of lead and was clearly fake!)

Isaac Newton was born in the same year Galileo died.[4] Legend has it that a 23-year-old Newton was looking at an apple tree in his estate in Woolsthorpe (the tree still stands) when he wondered if the force which pulled apples to the ground could also be responsible for holding the Moon in its orbit. Newton knew that the Danish astronomer Johannes Kepler had shown that the planets moved in elliptical orbits around the Sun. The speed of the planets as they orbited the Sun was such that a line connecting the planet to the Sun swept out an equal area in equal times:

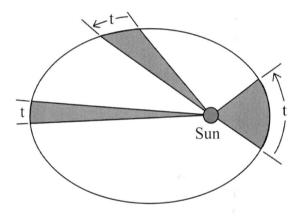

Newton wanted to analyse this motion, but found the mathematics of the day was not sufficient. As a result, Newton invented calculus (he actually invented calculus at the same speed at which it is normally taught in schools).

---

[4] And Albert Einstein was born in the same year James Clerk Maxwell died.

Using calculus, Newton could deduce the form of the law of gravity:

> *I began to think of gravity extending to the orb of the Moon, and having found out how to estimate the force with which a globe revolving within a sphere presses the surface of the sphere. From Kepler's rule of the periodical times of the planets, I deduced that the forces which keep the planets in their orbs must be reciprocally as the squares of their distances from the centres about which they revolve, and thereby compared the force requisite to keep the moon in her orb with the force of gravity at the surface of the Earth and found them to answer pretty nearly.*

So in 1687, Newton published the *Principia*, which has become generally recognised as one of the most important books ever published. In the *Principia*, Newton presented his law of universal gravitation for the first time. He stated that every mass in the universe attracted every other mass by means of the force of gravity. This gravitational force is proportional to the masses multiplied together, and inversely proportional to the square of the distance between the masses:

$$F = G \frac{m_1 m_2}{r^2}$$

where F is the gravitational force between two masses m_1 and m_2, r is the distance between the centres of those masses, and G is the *gravitational constant*. The value of the gravitational constant has to be found from experiment.

Newton's formula for universal gravitation is still used today for calculations involving masses which are not too huge. It was Newton who revealed gravity to the world.

However, there was an aspect of universal gravitation which left Newton dissatisfied. The formula seemed to imply that there was an instantaneous force exerted by one mass on the other – despite the fact that the two masses could be separated by a great distance over the vacuum of empty space. There was no indication of any intermediary substance between the masses which might be used to transmit the force. Newton would have been happy if one of the masses had pulled the other mass by a mechanical means, for example, but this "invisible hand" stretching out over vast distances with nothing in between made no sense. As Newton said in a letter in 1692:

> *That gravity should be innate, inherent and essential to matter, so that one body may act upon another at a distance thro' a vacuum, without the mediation of anything else, by and through which their action and force may be conveyed from one to another, is to me so great an absurdity, that I believe no man who has in philosophical matters a competent faculty of thinking, can ever fall into it. Gravity must be caused by an agent acting constantly according to certain laws; but whether this agent be material or immaterial, I have left to the consideration of my readers.*

This principle of Newton's "invisible hand" which can magically reach great distances in space is more generally called *action-at-a-distance*. Newton's instinct that action-at-a-distance was wrong eventually proved to be correct, but it took 200 years and Albert Einstein to provide an alternative.

General relativity

In 1907, when Einstein was sitting in his chair at the patent office in Bern, he had what he later called the "happiest thought of my life". Einstein imagined a person standing inside a stationary lift (or "elevator" in America) at the top of a tall building. Obviously such a person would feel the force of gravity. If, for example, the person dropped a ball then the ball would fall to the floor of the lift. But Einstein then imagined that the chain supporting the lift broke. As the lift fell many floors to the ground, the person inside the lift would not feel the force of gravity in any way. If, for example, the person tried to drop the ball again, the ball would not drop to the ground but instead it would just float in mid-air (relative to the person).

So Einstein realised that the force of gravity could be eliminated if the observer was accelerated in a particular way. This made Einstein believe that the force due to gravity and the force due to acceleration were equivalent. For example, if a person who is standing on the surface of the Earth (Figure a) in the diagram opposite) drops a ball, then the ball will drop to the ground under the force of gravity. This is different for an observer in a spaceship in deep space, far away from sources of gravity. If that observer drops a ball it will just float in mid-air. But if the spaceship is accelerated, as in Figure b) opposite, then a ball which is dropped will fall to the floor of the spaceship as if it is under the force of gravity:

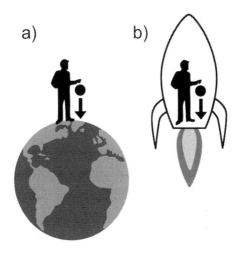

Einstein realised that the force due to gravity was **completely equivalent** to the force experienced during acceleration. This is called the *equivalence principle*. For example, a spaceship might generate its own artificial gravity by continuously rotating, rotation being a form of acceleration. NASA's proposed long-duration space exploration vehicle, NAUTILUS-X, includes a centrifuge to generate artificial gravity for its six-person crew:

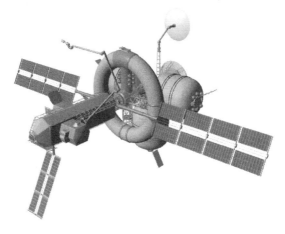

Remember that, according to the equivalence principle, the artificial gravity produced by the NAUTILUS-X centrifuge is **precisely equivalent** to conventional gravity. It is exactly the same thing.

Einstein then again imagined our spaceship in deep space undergoing acceleration. And he also imagined that there was a small window in the side of the spaceship through which light from a nearby star was entering the spaceship. According to the point-of-view of the occupant, the beam of light entering through the window appears to curve to the floor as the spaceship accelerates (as the light crosses the spaceship, it moves closer to the floor as the spaceship accelerates upward).

The following image shows this curved beam of light inside the spaceship:

So, the acceleration appears to curve light. But the equivalence principle states that this situation of the spaceship undergoing acceleration is equivalent to the spaceship under the influence of gravity. Hence, Einstein realised that gravity would also curve light. This was a radical thought: as light is massless it was assumed that it would not be affected by gravity.

Einstein realised that this prediction that gravity bent the path of light provided a way to test general relativity. He stated that light from the stars would bend as it passed the Sun. In a 1911 paper, Einstein proposed an ingenious experiment to test this bending of light: "As the stars in the parts of the sky near the Sun are visible during total eclipses of the Sun, this consequence of the theory may be observed. It would be a most desirable thing if astronomers would take up the question."

If we jump forward in time at this point, we find it was not until May 1919 that the British astrophysicist Arthur Eddington travelled to the African island of Principe in order to test Einstein's theory. Eddington was going to take photos of the Sun during a solar eclipse, which, as Einstein had suggested eight years earlier, was the only occasion when photographs of the stars around the Sun would be possible.

Unfortunately, the weather was not good, and for many days the cloud cover threatened to prevent a clear view of the eclipse. The weather was no better on the day of the eclipse, May 29th. However, the clouds broke for just a few minutes to allow Eddington to capture sixteen photographs of the eclipse. Two of Eddington's photos showed the positions of the stars clearly enough to confirm the bending of starlight around the Sun, thus providing the first experimental confirmation of Einstein's theory.

Eddington announced his result at a joint meeting of the Royal Society and the Royal Astronomical Society in London. As he was leaving the meeting, Eddington was famously asked if it was true that only three people in the world understood the theory of general relativity. When Eddington delayed in answering, the reporter said: "Don't be so modest, Eddington", to which Eddington replied: "Not at all. I am just wondering who the third person might be."

When the news of Eddington's confirmation reached Einstein, he was delighted but not surprised. Einstein noted that had the result been different, he "would have been sorry for the dear Lord, since the theory is correct."

So as gravity treats all objects equally – including light – Einstein realised that the type of the object is irrelevant for interacting with the gravitational force. Instead, the critical factor was the nature of space itself, as the motion of all moving objects depends on the underlying shape of space.

Einstein realised this could only be explained if space itself was curved by large masses. The trajectory of an object in free-fall, attempting to travel in a straight line in space, would then be curved towards mass.

The time-honored way of demonstrating this curvature of space is to consider a two-dimensional experiment: obtain a rubber-sheet and place a large mass in the middle so that it deforms the sheet. A small ball will then "orbit" the central mass as if it was a planet orbiting the Sun:

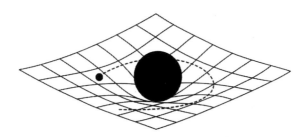

Another attraction of considering gravity as being the curvature of space was that it introduced the concept of the gravitational force being represented by a *field*, in the same way that other forces – such as the electromagnetic force – is carried by a field. The great attraction of the field approach is that it provided an alternative to the Newtonian action-at-a-distance interpretation of gravity.

Einstein was in agreement with Newton that action-at-a-distance was unacceptable, but for a different reason to Newton. Einstein realised that action-at-a-distance implied instantaneous transmission of information over great distances – something forbidden by Einstein's theory of special relativity which did not allow anything to travel faster than light. The only alternative to action-at-a-distance was provided by a field in space.

We are probably best acquainted with the idea of field being emitted from a magnet. If you place a magnet under a sheet of paper and sprinkle iron filings over the paper, the field lines become obvious:

So we can imagine a field as being an invisible entity which spreads through space, an entity which has a direction and strength which is specified for every point in space. Instead of action-at-a-distance, which implies Earth reaching out over thousands of miles of empty space to hold the Moon in orbit, under the field approach we now consider Earth as sending out a *gravitational field*. The direction of the Moon's travel is only affected by the strength and direction of the gravitational field in the immediate vicinity of the Moon.

The field approach eliminates the problems associated with action-at-a-distance. Instead of an instantaneous connection between two masses, any disturbance in the field takes time to travel between the two masses – it is no longer instantaneous. Special relativity is no longer contradicted as the rippling distortion in the field travels at the speed of light, and no faster. If the Sun was to disappear, for example, the Earth would continue to orbit the last position of the Sun for seven minutes until the bad news reached the inhabitants of Earth via the gravitational field. At that point, the elimination of the Sun's gravitational field would result in the Earth spinning-off into space.

After his great intuitive realisation of 1907 about gravity being the curvature of space, Einstein spent many years trying to represent this curvature in mathematical form. He soon realised the mathematics describing the curvature of four-dimensional spacetime was beyond him, so he enlisted the help of his former classmate at the Zurich Polytechnic, Marcel Grossmann (now a professor of mathematics), and the Italian mathematician Tullio Levi-Civita who had invented *tensors*, essentially a way of mathematically describing curvature of space.

In 1915, Einstein finally published the *general theory of relativity* which revealed how the curvature of spacetime depended on the distribution of mass and energy according to the following equation (don't worry about the details of the equation):

$$\underbrace{R_{\mu\nu} - \frac{1}{2} R\, g_{\mu\nu}}_{\text{The curvature of spacetime}} = \underbrace{\frac{8\pi G}{c^4} T_{\mu\nu}}_{\text{Mass-energy distribution}}$$

The left-hand side describes the curvature of spacetime and is called the *Einstein tensor*. The right-hand side describes the distribution of mass-energy and includes the *stress-energy tensor*. The equation as written in this form is deceptively simple. It hides the fact that this actually represents ten complex, non-linear equations that are very hard to solve (we will be considering one solution – the very first solution which considers the field around a spherical mass – in Chapter Five which is about black holes).

However, when Einstein wrote down this equation, he realised it had a problem. It appeared to indicate a universe which would contract and completely collapse due to the mass contained within the universe.[5] This was unacceptable to Einstein who – like most of his contemporaries – considered the universe to be a static structure which had existed forever. After all, the stars had shone in constant positions in the sky for the whole of recorded human history. The only solution Einstein could see was to introduce a new repulsive force which acted to counteract the contraction of the universe. So Einstein added a factor, Λ, which he called the *cosmological constant*:

$$R_{\mu\nu} - \frac{1}{2} R \, g_{\mu\nu} + \Lambda \, g_{\mu\nu} = \frac{8\pi G}{c^4} T_{\mu\nu}$$

↑ Cosmological constant

---

[5] It is interesting to note – though it is rarely mentioned – that the prediction that the universe should collapse in on itself was just as much a problem for Newtonian gravity as it was for general relativity. Newton was aware of the problem, but avoided it by saying the stars were so far away from each other that their gravitational pull was negligible.

It is interesting that Einstein placed his correction factor on the left-hand side of the equation which describes the curvature of spacetime. He might have chosen to place his correction on the other side of the equation, thus modifying the predicted energy of space. In fact, as we shall see later, this latter approach is the one favored in modern cosmology.

So Einstein went ahead and added the cosmological constant to his equation. However, in the 1920s, Edwin Hubble considered the redshifts of galaxies beyond our Milky Way and discovered the universe was actually expanding. On hearing this news, Einstein realised his blunder in horror. An expanding universe had no need of a cosmological constant to keep it static. Einstein realised he could have predicted the expansion of the universe from his equations before Hubble discovered it from observation, a prediction which would have stood as one of the greatest in the history of science. As a result, Einstein called his introduction of the cosmological constant the "biggest blunder" of his life.

However, as we shall see later, recent observations have revealed that the expansion of the universe is actually accelerating. This suggests the universe has a small but positive cosmological constant – so maybe Einstein was right after all!

Einstein's theory of general relativity remains our best theory of gravity – one hundred years after it was published. It has been tested many times, for example, to measure the deflection of light around the Sun, and it has always proved to be accurate.

So the general theory of relativity is undoubtedly accurate, and correctly models the curvature of spacetime. But there is nothing in the theory to explain **why** a large mass curves space. The equation captures the behaviour, but does not explain it.

But can we make an attempt to explain **why** spacetime is curved by mass? In other words, can we understand what gravity is, and why it exists at all. I believe we can, and in the later chapters of this book we will see why.

But for now we are going to move on to understand the effect of gravity on the universe as a whole.

3

COSMOLOGY

Cosmology is the study of the universe itself at the very largest scales. It considers the size and shape of the universe, the expansion of the universe, the beginning of the universe, and the eventual fate of the universe. It is distinct from astronomy which is the study of individual celestial bodies such as stars and planets, whereas cosmology is the study of the universe as a whole.

Cosmology does not focus on considering each individual star in the universe, which would be an impossible task for our current level of technology. Instead, cosmology operates by taking averages, for example, finding an average value for the density of the universe. Inevitably, this means cosmology is not a precise science, but fortunately the universe looks quite similar at very large scales so these approximations work surprisingly well.

It has to be said that until well into the 20th century, cosmology was not considered a proper science. It seemed to ask such profound questions – such as the origin of the universe – which did not seem to be the sort of questions which analytical science should be asking. Surely such questions should be the domain of metaphysics, and forever beyond the experimental reach of science? However, throughout the 20th century a series of observations about the origin of the universe and its expansion proved that these profound questions were not beyond scientific enquiry, and cosmology became established as a valid science.

The scale of the universe

Perhaps the most remarkable fact about the universe is its sheer scale. I do not think it is at all generally realised how extraordinarily vast the universe is. I believe this scale is important, and I would like to try to convey to you some feeling for this vastness. So in this section I will present some examples from astronomy.

I am sure you are aware that the Earth is one of eight planets orbiting the star that is the Sun. This forms the Solar System. But the Sun is just one star in our galaxy which is called the Milky Way (a galaxy is a collection of stars, gas, and dead stars such as black holes). The closest star to the Sun is Proxima Centauri which is 4.2 light years away, which means it would take light (which travels at the incredibly fast speed of 186,000 miles per second) 4.2 years to reach our nearest star.

It is a little bit hard to get our heads around the concept of a "light year" – it is not a measure of distance we use every day! Perhaps if I tell you that one light year is six million million miles you might get a better feel of the extreme distance involved.

Our galaxy is truly vast: 100,000 light years across. It is a spiral galaxy with four distinct arms (see the following diagram). The spirals are caused by the rotation of the galaxy:

COSMOLOGY

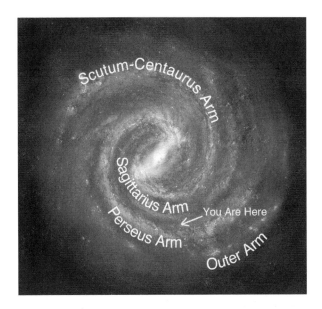

All the famous constellations of stars in the night sky are found within our galaxy. The Sagittarius constellation is located near the galactic centre (we will be considering the importance of the galactic centre in the chapter on black holes). The Solar System is located about two-thirds of the way out from the galactic centre.

There are many other fascinating structures to be found within our galaxy, including spectacularly beautiful giant clouds of gas called nebulas which may be as much as 100 light years in diameter. You may have seen beautiful pictures from the Hubble Space Telescope of the Eagle Nebula, located 7,000 light years away in the Sagittarius Arm, which includes the "Pillars of Creation" in which stars are formed (http://tinyurl.com/2wbzt9b).

So, how many stars do you think there are in our Milky Way galaxy? A hundred million? A billion? Well, there are actually about 200 billion stars in our galaxy! When you think how large the distance is between individual stars, I hope this gives you an impression of how vast our galaxy is.

Now let us consider other galaxies. The nearest galaxy to our galaxy is the Andromeda galaxy. This is also a spiral galaxy and is located about 2.5 million light years away. Galaxies attract each other and orbit each other just as planets orbit a star. This clearly shows how gravity is the dominant force in shaping the universe.

The Milky Way and the Andromeda galaxy are the two largest galaxies in a group of about 50 galaxies known as the *Local Group*. Andromeda – just like the Milky Way – has its own series of satellite galaxies which orbit it. The Andromeda galaxy made the news in 2012 when it was announced that the mutual gravitational attraction between it and the Milky Way was so strong that Andromeda will collide with the Milky Way in about 4 billion years time.

So here is the big question: how many galaxies do you think there are in the universe in total? A million? A hundred million? You might be astonished to hear that there are **over 100 billion galaxies in the universe!** When you think how large galaxies are, and how far apart they are, I hope this gives you a feel for the truly staggering scale of the universe. This is perhaps something which is not generally realised: the universe is really incomprehensibly huge.

And this will be one of the themes of the later chapters in this book. The universe is of such a completely different scale to any of the objects we encounter in our everyday lives. The difference is so vast that the universe effectively represents a completely different type of object. Might it be possible that the laws of physics – which we know so well at human scales – work differently at the scale of the universe itself?

All will be revealed later.

Olbers' paradox

The paradox I am going to describe now is interesting because it shows how much of the nature of the universe it is possible to deduce just by pure thought alone.

Several centuries ago, when hardly anything at all was known about the structure of the universe, several astronomers were puzzled why the night sky should be dark. At the time, the universe was assumed to be static and infinite. However, if there were an infinite number of stars in the sky, then any straight line drawn from your eye into the deepest depths of space would be bound to hit the surface of a star at some point. In this respect, the stars were like trees in a forest: if there were enough trees in the forest then it would be impossible to see through the forest to the other side – your line-of-sight would be bound to hit a tree at some point.

So if your line-of-sight was bound to hit a star at some point, then this would appear to indicate that the entire night sky should be covered with light, as there would be no point that was not covered by a star. The fact that the night sky is dark was therefore something of a mystery, and this apparent paradox is called *Olbers' paradox*.

The most obvious solution to the paradox was that there was not an infinite number of stars in the universe, which would appear to indicate that the universe was finite in size. This would represent a major revision of the model of the universe. So just by thinking logically about the problem, we can make discoveries about the universe from our armchair.

You might be tempted to raise various objections about Olbers' paradox which might occur to you. For example, you might say that the distant stars are so far away that their light would be so dim when it reached Earth it would be undetectable. Hence, the sky would be mainly dark. This is a strong objection. It is true that distant stars are more dim. However, it is also true that distant stars appear to have a smaller surface area to our eyes. Because of this, for a certain sized region of the night sky, it is possible to fit more dim stars into an area than it is possible to fit bright stars. The overall effect is that any area of the night sky should have exactly the same brightness.

Another objection you might raise is that the distant stars might appear dim due to interstellar dust obscuring the light. However, this would not work as the dust would increase in temperature and would then radiate the same light as the stars.

So Olbers' paradox was an early pointer to indicate that the model of an infinite and static universe was flawed. A new model of the universe was required.

COSMOLOGY

The Big Bang

Up until the 20th century the universe was assumed to be an eternal and unchanging structure: the so-called *Steady State* theory. However, our model of the universe changed when the world's largest telescope was completed in 1917.

The Hooker telescope is located in the Mount Wilson Observatory in the San Gabriel Mountains near Pasadena in California. At the time of its construction, it was the largest telescope in the world, a title it held for 30 years. A 4,000kg slab of glass was ground for five years to produce the 2.5 metre mirror.

The astronomer Edwin Hubble arrived at Mount Wilson in 1925 and used the telescope to examine the faint cloud-like objects which were believed to be nebulas. He revealed that these were not nebulas at all, but were in fact separate galaxies outside of our own galaxy.

Hubble went on to show that these galaxies were moving with respect to the Milky Way. The technique he used was discovered by Christian Doppler of Prague 100 years earlier. Doppler considered moving objects which emitted a sound of a constant wavelength. While it is true that the speed of the wave would not alter (we all know that the speed of light is a constant), the effect of the motion would be to "bunch up" the waves. In this way, an object moving toward an observer would appear to be emitting a sound of a higher pitch (the wavelength would be shorter), and an object moving away from an observer would appear to be emitting a sound of a lower pitch (the wavelength would be longer).

Doppler tested his theory on the flatlands of Holland, arranging for six trumpeters to ride outside a wagon on a steam train. The trumpeters emitted a constant middle C note. As the train passed the observers, the note was heard to rise and fall by a semitone. This change in wavelength is called the *Doppler shift*.

So it is possible to use the Doppler shift to determine if an object is moving towards you or away from you. Edwin Hubble examined the wavelengths of light from distant galaxies and found that they were shifted towards the longer, red end of the spectrum. Hence, the galaxies were said to have a *redshift*. This showed that the galaxies were moving away from us. This could mean only one thing: the universe was expanding.

From his observations, Hubble realised that the further away a galaxy was, the more it was redshifted. This meant that further objects were moving away from us at a faster speed. Hence, Hubble was able to express the relationship between the speed at which a galaxy was moving away, v, and its distance from us, d:

$$v = Hd$$

where H is called the *Hubble constant* (it is not really a constant – its value is known to have changed over time).

Now, here's a surprising thing. I am sure you are well aware of the law which says that nothing can travel faster than light. Well, strange as it may seem, it is actually possible for distant galaxies to be moving away from us faster than the speed of light. This is because the expansion of the universe is caused by the expansion of space itself (we will be examining this in more detail in the section on "The Shape of the Universe" later in this chapter). The cumulative effect of the expansion of space over vast distances means that distant galaxies can actually move away from us faster than the speed of light. The further apart galaxies are, the faster they are moving away from each other.

Bearing this in mind, we can obtain a value for the radius of the observable universe by considering the distance at which distant galaxies recede from us at a speed faster than the speed of light (and are hence invisible to us). If we put $v = c$ in the previous equation, we get a value for the distance:

$$R_H = \frac{c}{H}$$

where R_H is called the *Hubble radius*. We will be seeing this equation again later.

But if the universe is expanding, then that means if we look back in time then the universe was progressively smaller. And if we look far enough back in time we find that the universe must have emerged from an extremely small space. Hubble had discovered that the universe was not eternal and unchanging: the universe had a beginning. There must have been a single point of creation from which everything burst into existence. This event is called the *Big Bang*.

The Big Bang occurred 13.7 billion years ago. During the first moments of the Big Bang, the universe expanded from an extremely hot and dense point which contained all the matter and energy of the universe. Just imagine the mass of 100 billion galaxies squeezed into the size of a single atom. The temperature of the universe in these early stages was 1,000 trillion trillion degrees Celsius. At this temperature, everything melts. The tiniest building blocks of matter – quarks – swirled around freely in this particle soup called *quark-gluon plasma*.

However, after the first millisecond if its existence, the universe had cooled to a relatively chilly ten quadrillion degrees Celsius. At this lower temperature, three quarks could bond together to form protons or neutrons. Amazingly, every single proton and neutron in existence today was formed in this first millisecond after the Big Bang.

At this stage, the universe was opaque, meaning that photons of light could not travel freely but kept colliding with free protons and neutrons. Imagine you are at the bottom of a dark lake – you could not even see your hand in front of your face. That is what the opaque era must have been like.

However, when 380,000 years passed, the temperature had cooled to 10,000K. Electrons were now travelling slowly enough to be pulled into orbit around atomic nuclei composed of protons and neutrons. Hence, atoms were formed for the first time. This is known as the *recombination era*. Photons were now free to travel freely without being constantly scattered by protons and neutrons. So at this point the universe became transparent.

The photons which were released at this moment can be detected even today as they form the cosmic microwave background (CMB) radiation. This radiation has cooled over the last 13.7 billion years and it now has a temperature of only 2.7K. However, this radiation is everywhere, and is even responsible for 1% of the "snow" noise which you used to see on old analogue televisions. It is amazing to think that the interference on your TV screen was emitted 13.7 billion years ago! When it was discovered by accident in 1964, the CMB was the strongest evidence in support of the Big Bang theory.

When we look out into space with our most powerful telescopes, we are effectively looking back in time. This is because we are looking at objects whose light has taken longest to reach us. The Hubble Space Telescope is capable of producing distant images of galaxies in the earliest stages of formation. Eventually, at the furthest distance, we find we are looking at the point when the universe was opaque, so we can look no further back in time. At that point, we are looking at the CMB – the furthest (and earliest) thing we can ever see.

COSMOLOGY

The Wilkinson Microwave Anisotropy Probe (WMAP) was launched in 2001. It is located far enough away from the Earth to avoid the magnetic field or any other source of interference. Temperature measurements taken by WMAP of the CMB revealed it to be relatively smooth in all directions, with only small clumps and ripples (see following image).

These clumps and ripples at the dawn of time were extremely important, though, because they formed the seeds of the present-day structure of the universe. The gravitational attraction of the clumps pulled-in more mass, so the clumps grew to ever larger size, eventually becoming stars and galaxies.

Intriguingly, a paper was published in 2010 which revealed there was one remarkable feature clearly visible on the CMB.[6] As the paper's authors explain:

> *Shortly after the WMAP sky maps became available, one of the authors noted that the initials of Stephen Hawking appear in the temperature map. Both the "S" and "H" are beautifully vertical in Galactic coordinates, spaced consistently. We pose the question, what is the probability of this occurrence?*

If we examine the image of the CMB, Stephen Hawking's initials can indeed be quite clearly seen (inside the white ellipse):

---

[6] C.L. Bennett et. al., *Seven-Year WMAP Observations: Are There Cosmic Microwave Background Anomalies?*
http://arxiv.org/abs/1001.4758

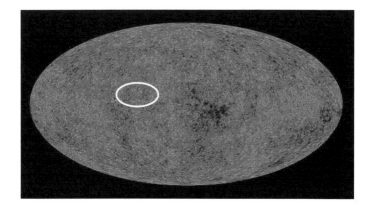

More recent results, most notably from the Planck space observatory in 2013, no longer show the Hawking initials, but other unexpected anomalies have been revealed.

The image below was produced from the Planck observatory and has two anomalous features enhanced. Firstly, the curved line shows asymmetry in the average temperatures revealing that the sky is slightly warmer in the south than in the north. This is the wonderfully-named "Axis of Evil". There is also a large cold spot (shown inside the ellipse). This has been called the Great Void of Eridanus as it is in the direction of the Eridanus constellation:

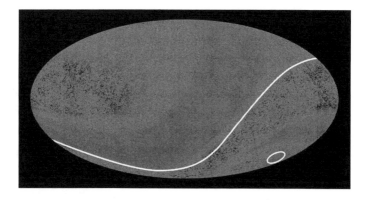

These are not the only anomalies to have been discovered recently. Only relatively recently has it became possible to map the positions of galaxies and galactic clusters in three dimensions. This has revealed peculiarities in their motion and positioning. Galaxies have a tendency to clump together in clusters and filaments, leaving large voids which are free of galaxies. This clustering introduces gravitational forces which pull galaxies out of their expected trajectories.

As was discussed earlier, our Milky Way galaxy is one of a number of galaxies which form the Local Group. It has been discovered that the Local Group is travelling at about 400 miles per second in the direction of the Virgo cluster of galaxies. This area to which so many galaxies are attracted is called the Great Attractor. The reason for this mass movement is unknown as the gravitational pull of the Virgo cluster is not powerful enough. This migration of galaxies has been called *dark flow*.

No one knows the reason for these anomalies. However, it is clear that the universe is not as smooth and featureless as had been believed.

The shape of the universe

In our discussion of gravity in the previous chapter, it was explained how the presence of mass in general relativity curves space. Well, the universe obviously contains mass, so the universe must logically have some form of curvature. So it appears we can talk about the "shape" of the overall universe. What shape is the universe?

You might find it a bit strange talking about the "shape" of the universe, because you might think the universe doesn't have a shape – it just goes on forever, doesn't it? Surely it is inconceivable that the universe has an edge, like a wall, or a drop at the end which you fall off if you go too far? Well, maybe it is the case that the universe is infinite in size, but, surprisingly, it is also possible for the universe to the finite in size but still have no edge. Which situation is true depends on the shape of the universe.

If the density of mass in the universe is high enough then space can become so curved that it loops around on itself. Imagine space looped around in a circle as if it is the inside of a ping pong ball. You could travel forever around the inside of the ball but never reach an edge. It has also been called the "Pac Man" universe: you go out of one side of the game and come back in on the opposite side. This is the situation when the universe is finite in size but has no edge. This is called a *closed* universe. Imagine the universe as a "ball" of space.

The expansion of the universe is a unique form of motion. Normally, motion involves the movement of an object against a pre-existing background of space. But when the universe expands, it is not a case of the farthest galaxies flying away through pre-existing space. After all, there is nothing outside the universe, so that means there is no pre-existing space outside the universe, so what could the universe possibly expand **into**? No, what is actually expanding is this "ball" of space itself.

There are two other possibilities for the shape of the universe, depending on the density value. These possibilities are illustrated in the following images of three equilateral triangles drawn in the three different types of space:

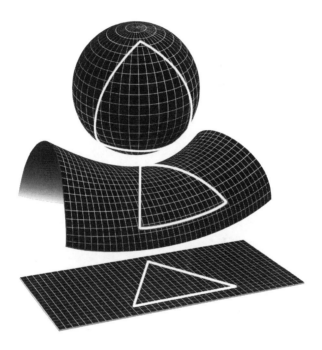

The first example shows a closed universe, as if it was the inside of a ping pong ball. In this case, parallel lines eventually meet, and the sum of the angles of a triangle exceeds 180 degrees (although this would only be noticeable on vast triangles of cosmological scale).

The second example shows the case when the density is sufficiently small. This is called an *open* universe. Instead of space being closed, space diverges in an open universe as if it is drawn on a saddle. In this case, parallel lines eventually diverge, and the sum of the angles of a triangle is smaller than 180 degrees.

The last example shows a *flat* universe. In this case, the value of the density is a particular value so that space is neither closed nor divergent. The sum of the angles of a triangle is equal to 180 degrees.

Let us now try to analyze these three scenarios in more depth. At the end of the previous chapter we were introduced to Einstein's equation for general relativity, and it was explained how the equation is very difficult to solve. In fact, exact solutions have only been found when simplifying approximations have been made. For the case of the entire universe, the first exact solution was developed by the Russian physicist Alexander Friedmann in 1922. The solution was only possible if we make the simplifying assumption that the universe has a very similar consistency and structure throughout its entire volume. We say that the universe must be *isotropic*, meaning that the universe looks the same in all directions – no matter where you stand in the universe.[7]

Of course, you might argue that the universe is far from isotropic: there are different galaxies in different directions. However, with there being over 100 billion galaxies in the universe, the distribution of those galaxies is very equal, and the resultant texture of the universe at the largest scales is

---

[7] Sometimes it is said that the universe must also be *homogeneous*, meaning that it must be the same all over. However, an isotropic universe is necessarily also homogeneous.

very smooth. The universe does indeed appear to be isotropic.

The following equation is the first of the two Friedmann equations. The equation is derived from Einstein's equation. It relates the curvature of the universe, k, to the density of the universe, ρ. So it can be used to derive the curvature of the universe if we know the density of the mass in the universe:

$$H^2 = \frac{8\pi G}{3}\rho - \frac{kc^2}{a^2}$$

In the equation, we have already met the Hubble constant, H. The only other term you might not recognise is the scale factor, a, which basically says how large the universe currently is. We won't be seeing this term again, so don't worry about it. You should be able to recognise every other term (including the gravitational constant, G, and the Hubble constant, H).

We would like to calculate the vitally-important density of a flat universe. This is because a flat universe is perfectly balanced between being open and closed. A flat universe has zero curvature, so to derive the equation for that universe we set the curvature, k, to zero in the previous Friedmann equation. When you multiply something by zero, that term disappears, so we are left with:

$$H^2 = \frac{8\pi G}{3}\rho$$

and if we reorganize the terms in this equation we get a formula for the density of a flat universe:

$$\rho = \frac{3H^2}{8\pi G}$$

This density value for a flat universe is perfectly balanced between being an open universe and being a closed universe. Hence, this is called the *critical density*. If the actual density of the universe is more than this critical density then the universe is closed. If the actual density of the universe is less than this critical density then the universe is open.

We will be seeing this vital equation again later.

But a flat universe contains matter, so how is it possible that a flat universe is not curved? Well, this is possible because the curvature of space (which pulls the universe inward) is perfectly balanced by the energy of the outward expansion of space. In the previous equation for the critical density, we can therefore see that the density depends on the rate of expansion of the universe described by the Hubble constant, H. If the universe is expanding faster, the critical density is larger.

So in a closed universe, the density of the universe is greater than the expansion energy given by the Hubble constant. This means that gravity will tend to pull the universe back together again. In the long run, this means that a closed universe is bound to eventually collapse. Even if a closed universe is currently expanding, gravity is bound to eventually win the day: the expansion will slow and eventually reverse, and the universe will collapse.

This is not the case in a flat or open universe. As was described earlier, in a flat universe the gravitational pull is perfectly balanced by the expansion energy: the expansion of the universe will slow down, but it will never collapse. In an open universe, the expansion energy is much greater and the universe keeps on expanding forever.

So the actual density of the universe – and its relation to the critical density – determines the eventual fate of the universe.

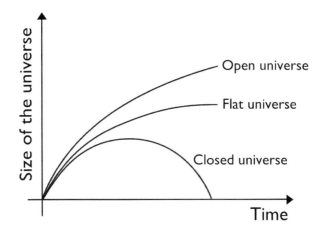

So is the universe actually closed, open, or flat? The evidence is obtained from the CMB radiation images which we considered in the previous section. The curvature of the universe makes it act like a magnifying glass, bending light and making distant objects either larger or smaller. If the universe was closed, distant clumps in the CMB would appear larger. If the universe was open, the clumps in the CMB would appear smaller. The fact that the clumps appear to be precisely the predicted size has resulted in NASA recently announcing: "We now know that the universe is flat with only a 0.5% margin of error."

This is really quite a remarkable result, and very unlikely to happen by chance. It requires the density of mass-energy of the universe to be set extremely precisely. If the density was slightly greater then the universe would have rapidly collapsed into a "Big Crunch". If the density was slightly less then the universe would have expanded too quickly for complex structures such as galaxies and stars to form.

So what could possibly be the reason behind this crucial flatness of the universe? One theory thinks it has a solution …

Inflation

We have just seen that a flat universe is essential to the development of interesting structures such as stars and galaxies (and therefore humans!). But this poses something of a mystery. Why should the universe be flat? Why is the density of the universe apparently fine-tuned to such a precise value? Why is the universe such an interesting place to live in, rather than just collapsing or flying apart? This is called the *flatness problem*.

In 1981, the American physicist Alan Guth believed he had found a solution. He proposed that in a tiny fraction of a second after the Big Bang, the universe expanded at an incredible rate. In just one hundred million trillion trillionth of a second the universe expanded from about the size of a subatomic particle to the size of a baseball. This period of incredibly rapid expansion is called *inflation*.

(It is proposed that the reason for this inflation is the negative pressure of the vacuum in the universe at that early stage. I won't go into detail about this at this point because we will be considering negative pressure in the next chapter.)

After this period of inflation, the universe is then predicted to expand at the normal, much lower expansion rate. But the period of inflation was enough to expand the universe to a much larger size than that predicted by the conventional Big Bang theory.

So how does this solve the flatness problem? Well, inflation works by smoothing and flattening the universe. If you imagine you have a rubber sheet full of ripples, as you stretch the sheet to a much larger size, the sheet becomes flatter and the ripples disappear. This is the effect inflation has on the universe.

COSMOLOGY

If you are standing on a bowling ball, the curvature of the ball will be very obvious to you. However, if you expand that ball up to the size of the universe, then, at a local level, the ball will appear completely flat. The same principle applies to a universe that has undergone inflation. Inflation is proposed to have expanded the universe to a vastly larger size than the observable universe (at least a hundred sextillion times larger[8]). This means the part of the universe we can actually see is just the tiniest fraction of the entire universe. So, just like the blown-up bowling ball, our observable universe appears flat.

There is another reason why inflation is an attractive hypothesis. Let us consider two galaxies, positioned either side of the Earth, and both of them 9 billion light years from Earth. This means the distance between the galaxies is 18 billion light years. As the universe is only 13.7 billion years old, this means there has not been enough time for light to pass between the two galaxies. Essentially, they are isolated from each other. But when we look around the universe, we find it looks remarkably uniform, and the CMB temperature of all areas of the universe is remarkably similar. This is surprising because it appears that the temperature has had a chance to equalise, ironing-out any differences. However, because it is not possible for light to have travelled between the galaxies, it should not have been possible for this equalisation to occur. Heat could not have flowed from warmer regions of the universe to the cooler regions. We would expect to see a universe with far less uniformity than we observe. This is called the *horizon problem*.

Inflation presents a possible solution to the horizon problem because, once again, the stretching of the universe

---

[8] Figure taken from *The Inflationary Universe* by Alan H. Guth.

during the inflationary period has the effect of smoothing the universe. According to inflation, our observable universe represents only a small fraction of the total universe. Therefore, the region which has become our observable universe was actually a very small volume of the universe before inflation occurred, and the objects in that small volume could all communicate with each other. There was enough time for heat to flow from the warm regions to the cold regions and smooth-out any differences.

The inflation hypothesis has been widely accepted and has been incorporated into what is known as the "standard model of cosmology" which, like the standard model of particle physics, has about 20 free parameters (numbers which cannot be derived but have to be obtained by observation). However, the standard model of cosmology has nowhere near the same level of experimental verification. I recently read that the standard model of cosmology is a "triumph of modern physics". I am not convinced. I would describe it more as "a list of things we do not yet understand". It is also the case that, thirty years after it was first proposed, serious cracks are now starting to appear in the model of inflation.

In a 2011 *Scientific American* article, Paul Steinhardt – one of the developers of the inflation hypothesis – asked "Is the theory at the heart of modern cosmology deeply flawed?" (see http://tinyurl.com/cfnhwkw). According to Steinhardt, inflation would have to be fine-tuned to produce the smooth properties of the universe we observe today. Another strong criticism is that inflation should keep occurring throughout our universe: it is very easy to start inflation, but it is very difficult to stop it. Bubble universes are predicted to keep appearing in space. In fact, inflation predicts an infinite number of bubble universes will appear, with each bubble universe having different properties: some smooth, some highly non-uniform. So inflation is not such a predictive theory as had been assumed. A theory which predicts everything predicts nothing.

The simplistic original theory of inflation as proposed in the early eighties has been replaced by a theory which lacks predictive power, and seems fundamentally flawed. The opportunity is there for a new theory. In the later chapters of this book, a new, simple hypothesis will be presented for producing a flat universe.

4

DARK MATTER AND DARK ENERGY

The recent discovery of the Higgs boson represents the last piece of the jigsaw of the standard model of particle physics. The standard model was formulated in the late 1970s and it is one of the greatest achievements in the history of physics. It describes all of the known particles and their interactions. However, because the standard model does not describe gravity, it has been called the "theory of **almost** everything". Unfortunately, we have recently discovered that it is more like the theory of not very much at all.

We have discovered that the standard model only describes about 5% of the material of the universe.

To put it another way, we simply have no idea what the vast majority of the universe – 95% – is made of. In this chapter we will examine the mystery of the missing 95%. We will see that there are two components to the missing substance, and these are known as dark matter and dark energy.

This is what the universe is really made of.

Dark matter

In 1933, the Swiss astronomer Fritz Zwicky considered the motion of galaxies in the Coma Cluster (a large cluster which contains about 1,000 galaxies). He estimated the mass of the cluster by measuring its brightness, and, based on that mass, he predicted the motion of the galaxies on the edge of the cluster. Zwicky discovered that much of the required mass appeared to be missing. Though the nature of this missing mass was completely unknown, Zwicky gave the German name "dunkle materie" to the material, which translates to "dark matter". It is now believed that 90% of the mass of the Coma Cluster is in the form of dark matter.

Unfortunately, Zwicky was a difficult man to get along with (he was dubbed "borderline psychopathic" in a BBC documentary *Most Of Our Universe Is Missing*, available on YouTube). Zwicky referred to his colleagues as "spherical bastards" – because they were bastards no matter which way you looked at them. As a result, Zwicky's work was ignored for many decades.

However, in the 1970s, Vera Rubin of the Carnegie Institution of Washington had access to more accurate equipment and she used it to measure the speed of rotation of stars lying on the edge of spiral galaxies. She started by examining the Andromeda galaxy which, as was described in the previous chapter, is the nearest spiral galaxy to our own galaxy.

If you consider our solar system as an example, the outer planets orbit the Sun slower than the planets which are nearer the Sun. For example, Neptune, the farthest planet, takes 164 Earth years to orbit the Sun. And for galaxies as well, this is what we would expect. We would expect the stars on the edge of the galaxy to orbit the galactic centre at a slower speed than the stars nearer the centre.

However, this is not what Vera Rubin discovered. She discovered that the stars on the edge of the galaxy orbited the galactic centre in the same time as the stars nearer the centre. This meant the outer stars were travelling much faster than expected. Put simply, she discovered that the stars were orbiting too fast: the gravitational pull of the galaxy should not have been sufficient to hold the stars in orbit. Rubin's calculations revealed that galaxies must contain ten times more matter than is held in the visible stars. Eventually, other astronomers corroborated Rubin's result and it soon became accepted that most of the material in galaxies was dark matter. It appears that each galaxy resides inside a sphere of dark matter, and the radius of that sphere is approximately ten times the radius of the galaxy.

So just what is this mysterious dark matter? The first possibility is that it is just conventional matter, i.e., atoms composed of protons and neutrons. This would have to be in a form so that it does not emit any light (and is therefore "dark" and invisible). This might include black holes or very dense neutron stars. These type of bodies which are proposed to constitute dark matter are called MACHOs (Massive Astrophysical Compact Halo Object). However, MACHOs seem unlikely to be the solution to dark matter as they tend to be isolated objects, whereas it is known that dark matter is spread fairly evenly over a galaxy.

The opposite of being MACHO is to be a WIMP, so the other possibility is that dark matter is composed of WIMPs (Weakly Interactive Massive Particles). These would be new particles, not contained in the standard model. These particles would have to be much more massive than protons, but would have to interact only minimally with ordinary matter – which makes them very difficult to detect. They could not radiate light or scatter it, so they could have no electric charge.

Several experiments to detect WIMPs are based deep underground in converted mine workings and tunnels such as the Soudan mine in Minnesota, the Homestake mine in South Dakota, and the Boulby mine in the north of England. The thinking behind these experiments is that most particles from space will interact with conventional matter in the earth and will not reach deep down through the rock – only the non-interacting WIMPs will fly straight through the earth to the detector (of course, this raises the problem of how you manage to stop the WIMPs flying unhindered through your detector).

Another experiment to detect dark matter particles is the Alpha Magnetic Spectrometer (AMS), the most sophisticated (and expensive!) particle detector ever launched into space. It cost a cool two billion dollars, and orbits the Earth attached to the International Space Station. It contains a giant, specially-designed magnet that bends the paths of particles that pass through the detector, revealing information about their mass, energy, and velocity. About 1,000 cosmic rays are recorded per second, and this data is compressed and sent down for analysis at CERN.

Recent results from the AMS appears to show evidence that dark matter might be composed of a hypothetical WIMP called a *neutralino*. If you consider the predicted mass of a neutralino, and multiply that by the number of neutralinos that are supposed to exist, the result just happens to match the observed amount of dark matter. So where do neutralinos come from? There is a hypothesis in particle physics called *supersymmetry* in which each particle has a paired symmetric particle. For example, the photon is paired with the photino, and the Higgs boson is paired with the Higgsino. The neutralino is another particle predicted by supersymmetry (surprisingly, not paired with the neutrino). It is theorised that positrons should be emitted when neutralinos annihilate each other. As the AMS has detected an excess of positrons, this is taken as evidence that neutralinos might form dark matter. However, these particles could also be produced from other high-energy sources such as pulsars: rotating neutron stars. Hence, the evidence for WIMP dark matter is still not convincing and there have been accusations that CERN has been hyping the results for the media.

It remains possible that there is yet another possible solution to the dark matter conundrum, and that is that dark matter might not exist at all ...

MOND

Mordehai Milgrom of the Weizmann Institute in Israel is not a fan of dark matter: "People who deal with dark matter will not stop by themselves – they will stop when the money stops. And this is what they will want to continue to do as long as they are given the chance and the resources to do it. And the fact that they do not detect dark matter will just mean to them that dark matter is very difficult to detect."

Milgrom believes that dark matter is not necessary to explain the flat rotation curves of galaxies. The extra gravity required can come instead from a modification to Newtonian gravity. Milgrom noted that stars on the edge of galaxies do not have to accelerate as much as stars near the galactic centre (this is because their radius of orbit is less curved). Milgrom's suggested modification to gravity is that an object which is undergoing less acceleration would experience a greater pull of gravity. This meant that stars would orbit at constant velocities, no matter their distance from the galactic centre. This theory was given the name MOND (Modified Newtonian Dynamics).

MOND has been criticised for violating the law of conservation of momentum – not something you really want in your theory. Also, recent observations of gravitational lensing (whereby light is seen to bend around galaxies) have lent further weight to the theory that dark matter really exists in its own right. The galactic cluster known as the Bullet cluster was formed when two clusters collided. As a result of this tremendous collision, measurements of gravitational lensing has revealed that the dark matter has been separated from the main mass of galaxies. This result cannot be explained by a modified theory of gravity such as MOND.

So dark matter remains a mystery for the time being. But what about the other puzzling constituent of the universe ...

Dark energy

As discussed in the previous chapter, it is known that the universe is very close to spatially flat. This spatial flatness is achieved when the mass-energy density of the universe is equal to the critical density (the formula for which we derived in the previous chapter). However, this raises a problem as the total amount of matter (including dark matter) is measured to be only 30% of the amount required to bring the density up to this critical density. As Ta-Pei Cheng says in his excellent textbook *Relativity, Gravitation, and Cosmology*: "The Friedmann equation requires a flat universe to have a mass-energy density exactly equal to the critical density. Yet observationally, including dark matter, we can only find less than a third of this value. Thus it appears that to have a flat universe we would have to solve a missing energy problem."

What could possibly account for the missing mass-energy? Even taking dark matter into account, it appears that a further 70% of our universe is missing!

So what is this missing 70%? It is generally stated that there must be a vast amount of energy in space which we cannot detect, and even though we cannot detect it we know precisely the amount of this missing energy which exists: it is the amount required to raise the mass of the universe up to the critical density. So it is proposed that there is another invisible substance known as *dark energy* which pervades space, and makes up this missing 70% of the total mass-energy of the universe.

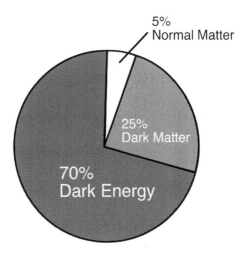

Now, at this point, I must admit that I have to raise an objection to this reasoning. As we discussed in the previous chapter, inflation is supposed to stretch the universe so that the visible universe is only a small part of the true size of the universe. This guarantees that the visible universe is flat (or very close to flat). As shown in the following diagram, the vast extent of the universe outside the visible universe might well have significant curvature, but the much smaller visible universe is guaranteed to be flat. In fact, the entire universe could have absolutely **any** type of curvature, but the visible universe would still appear to be flat.

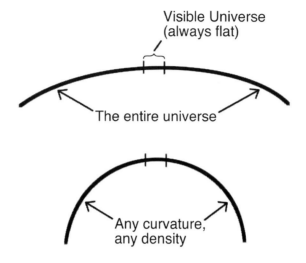

Bearing in mind the fact that the curvature of the universe is dependent on its density, this means that the universe in its entirety could have absolutely any value for its density but the visible universe would still always appear flat. If you have a very large volume of space at a constant density, then a very small region is always going to tend to be spatially flat – no matter what the density.

So, here is the crucial point. How is it possible to infer **anything** about the density of the entire universe just by observing the flatness of the visible universe? Inflation **guarantees** that the visible universe will appear flat, even if the entire universe does not possess the critical density for a flat universe. And, assuming a homogeneous universe, the density of the visible universe is going to be just the same as the density of the entire universe.

So how can it be valid to observe the flatness of the visible universe and then infer that dark energy is required to achieve this flatness? No matter what the density of the entire universe, no dark energy is required to achieve flatness in the visible universe. The reasoning seems flawed.

This value of 70% for the amount of dark energy in the universe is very well publicised. The reasoning I am presenting here appears to cast doubt on the accuracy of this figure. I have asked several cosmologists about this and I have not yet received a convincing answer to my objection.

However, there is yet another reason why the dark energy hypothesis is popular. And that is because of a truly remarkable recent result ...

The accelerating expansion

In the 1990s it had been known for several decades that the universe began with the rapid expansion of the Big Bang. It was also believed that in the billions of years since the Big Bang, the sum total of mass and energy in the universe would have been attracted together due to gravity, and this would inevitably have the effect of slowing the expansion of the universe. So in the 1990s, two competing groups set out to determine the expected rate of deceleration of the universe. One group was the Supernova Cosmology Project based at the Lawrence Berkeley National Laboratory in California, and was headed by Saul Perlmutter. The other group was the High-Z Supernova Search based at the Mount Stromlo Observatory in Australia, and was headed by Brian Schmidt and Adam Riess. But how could it be possible for these teams to measure how much the expansion rate was slowing down?

The only way to do this would be to consider extremely distant objects, over 5 billion light years away, which would also have to be extremely bright so that their emitted light would be visible. What was needed was an object whose brightness was known and was constant. These objects which all have the same known brightness are called *standard candles*.

One type of celestial body perfectly fitted the role of a standard candle. A supernova is a star which has burnt all its fuel and undergoes a thermonuclear explosion, emitting a tremendous amount of energy (we will be encountering supernovas again in the next chapter on black holes). There are different types of supernovas. The particular type of supernova we are interested in is called a Type 1a supernova. These occur in white dwarf stars, which are old stars which are small and incredibly dense. These stars are stable as long as their mass remains below 1.4 solar masses (known as the Chandrasekhar limit). If they attain any additional mass, maybe from neighbouring stars, these stars undergo a nuclear explosion, exploding in just ten seconds, and releasing a tremendous amount of energy. Because these white dwarfs all have the same amount of mass before they explode – 1.4 solar masses – the resultant supernova explosions all have the same brightness. This means that Type 1a supernovas make perfect standard candles for measuring the expansion of the universe.

Unfortunately, it is very difficult to find supernovas. They only burn brightly for three weeks, and in any given galaxy a supernova will explode without warning only once every 300 years. However, all is not bad news. With billions of galaxies in the observable universe, there are dozens of supernovas every night. So the international team had an ingenious idea. They took snapshots of thousands of galaxies, and then repeated those snapshots three weeks later. If any object in the second set of photos was considerably brighter then it could be recognised as a supernova. By scanning the entire sky in this method, the team were assured of a regular supply of supernovas. Perlmutter and his team dubbed this method "Supernovas on Demand".

The supernovas were then observed by the Keck telescope in Hawaii, and also by the Hubble Space Telescope. Because the brightness of the supernovas was precisely known, it was possible to determine how far away they were by measuring that brightness. So the distance to a supernova was determined by its brightness, and the speed at which it was moving away was determined by its redshift.

When the team plotted their results, they were puzzled. They found the plots of their supernovas deviated from the usual linear Hubble curve, but in a way which suggested that the expansion of the universe was accelerating – not decelerating! The graph showed that the expansion rate of the universe was smaller in the past, so that an object at a certain distance was not receding as fast as would be expected. The diagram below is typical of the sort of supernova plots (the black dots) they examined:

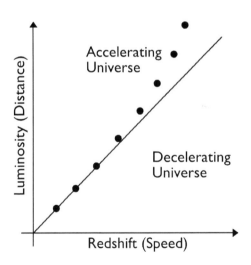

If you remember back to our discussion about the expansion of the universe in the previous chapter, you will recall that none of the available scenarios (open, closed, or flat universe) led to an accelerating expansion. In fact, in all three cases the rate of expansion of the universe was predicted to slow down. So this result came as a complete shock to the cosmology community. Saul Perlmutter, Brian Schmidt, and Adam Riess received the Nobel prize for this extraordinary discovery.

So what could possibly be causing this accelerated expansion of the universe? Intriguingly, a perfect candidate is Einstein's cosmological constant which we considered back in Chapter Two. If you remember, Einstein introduced the cosmological constant to prevent the collapse of the universe which was predicted by general relativity. The cosmological constant introduced a form of anti-gravity force which could hold the universe in a steady state. It was only when Edwin Hubble discovered the universe was expanding that Einstein realised his blunder.

Well, the cosmological constant can potentially do more than just hold the universe in a steady state: its anti-gravity force could be powering the expansion of the universe. But just what is this cosmological constant? Well, it might be the case that the cosmological constant is just another mathematical constant, with no obvious physical cause. In other words, maybe Einstein was correct and we should adjust the form of his equation to include this additional constant. But physicists seem reluctant to modify the equation for general relativity, and would rather seek some additional physical cause which plays the same role as the cosmological constant.

It is currently believed that dark energy might be nothing more than the energy of empty space, otherwise known as the *vacuum energy*. That might sound very strange: how on earth can empty space have an energy? Well, it turns out that empty space is not so empty after all, and that is all due to the effect of quantum mechanics.

In quantum mechanics, the Heisenberg uncertainty principle says that there will always be a degree of uncertainty in how accurately we can know a combination of certain particle properties. You may have heard that the more accurately we determine the position of a particle, the less accurately we can know its momentum. For example, if we arrange for a particle to hit a screen, we can determine the particle's position with 100% accuracy, but we will have no way of knowing how fast the particle was travelling when it hit the screen – there is maximum uncertainty about its momentum.

There is a similar uncertainty about a particle's time and energy. This means there is a fundamental limit on how accurately we can measure the energy of a system at a given moment in time. Incredibly, this means that for a short period of time, there can be energy produced in empty space in which all the matter has been removed. Even more incredibly, according to the mass-energy equivalence described by $E = mc^2$, this vacuum energy can be enough to produce a particle! It is as though the energy is "borrowed" from the vacuum. These so-called *virtual particles* can appear out of nothing.

These virtual particles can only exist for a very short period of time before the energy debt has to be paid back. Very shortly after the particles appear, they annihilate each other and disappear again. However, for the period that the particles are in existence they generate a very small amount of energy in empty space: the vacuum energy, also known as *zero-point energy*.

You might find it hard to believe that empty space can contain energy. However, this has been experimentally proved by measuring the *Casimir effect*. The Casimir effect is quite a remarkable experiment in which two metal plates are placed very close to each other – just a few micrometres apart. Virtual particles pop in-and-out of existence between the plates, but because of the very small distance between the plates this introduces a limitation on the type of particles which can be produced. As a result, fewer virtual particles are produced between the plates than outside the plates. The resultant difference in pressure generates a small force which pushes the plates together. This force has been measured: virtual particles are a reality.

Interestingly, if you have two ships floating next to each other, they will gradually come together for a very similar reason to the Casimir effect. This is because only waves of a limited number of wavelengths can fit between the ships – the wavelengths of the waves outside the ships being unlimited. As a result of this inequality, there is a small pressure which forces the ships together.

Of course, as you might have predicted, the idea of empty space producing energy has attracted the attention of many inventors who believe they can harness this energy to produce power. Unfortunately, all their efforts are doomed to failure. Zero-point energy is the lowest amount of energy that can exist. If you try to draw-off that energy from a region of empty space – to produce warmth, for example – you will only end up heating the empty space with your own equipment as the energy of the vacuum is guaranteed to be less than the energy of any equipment you might attach to it.

Vacuum energy has a particular property which makes it very interesting from the point of view of explaining the accelerated expansion of the universe, and that property is *negative pressure*. If you consider the cylinder filled with a vacuum in the diagram opposite, it can be seen that the vacuum will inevitably try to pull the piston further into the cylinder (this is inevitable as the vacuum has the lowest possible energy). In this sense, we can consider the vacuum as having negative pressure (if the vacuum had positive pressure then it would be forcing the piston out of the cylinder):

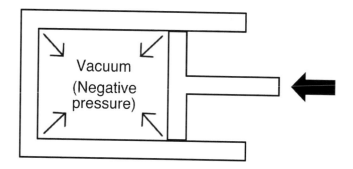

Now let us consider a normal gas under normal positive pressure in the cylinder. The gas would tend to push the piston out of the cylinder (as might happen in a steam engine, for example). In which case, the gas has done work, and the energy contained in the gas is **reduced** as a result.

However, if we now consider our cylinder full of vacuum, as we have seen, the vacuum will tend to pull the piston in. If the piston moves out of the cylinder instead, then the energy contained in the vacuum is **increased** as a result (the opposite to the case of the gas with positive pressure).

So the vacuum energy can result in negative pressure. And this is very interesting, because pressure is a form of energy. General relativity predicts that not only is mass gravitationally attractive, but positive pressure is also gravitationally attractive. And so, quite incredibly, negative pressure is predicted to result in gravitational repulsion! As Alan Guth says in *The Inflationary Universe*: "According to general relativity, a positive pressure creates an attractive gravitational field, as one might guess. A negative pressure, however, creates a repulsive gravitational field."

So next time someone says that gravity is a force which is always attractive, you can correct them: negative pressure results in gravitational repulsion!

And this is why the vacuum energy is a candidate for the mysterious dark energy: the repulsive gravitational effects of negative pressure could be powering the expansion of the universe.

You might be wondering why the gravitational attraction due to the energy of the vacuum does not negate the gravitational repulsion due to the negative pressure. After all, Einstein showed that energy is equivalent to mass, so surely the vacuum energy should be gravitationally attractive. The answer is, yes, the vacuum energy **is** gravitationally attractive, but this does not completely cancel-out the gravitational repulsion. This is because the mass (attractive) and pressure (repulsive) elements affect the accelerated expansion of the universe differently.

In order to see this, we need to consider the second of the two Friedmann equations. This equation provides a formula for the acceleration of the expansion of the universe, which is given by:

$$-\frac{4\pi G}{3}(\rho + 3P)$$

Where ρ is the density (which we have seen before), and P is the pressure. We can see that the pressure term is multiplied by three. This is because, very unusually, all observers must measure the same energy and pressure of the vacuum – no matter how they are moving. As a result, this value of three is there purely because there are three dimensions of space![9] The repulsive component of the vacuum energy, P, is therefore three times stronger than its attractive element, ρ. So the second Friedmann equation predicts an accelerating expansion.

But this all depends on the actual value of P. And that is where the problems really start ...

The biggest error in science

The required value for the vacuum energy is extraordinarily small – in the order of a millionth of a billionth of a joule per cubic centimetre. The effect has a cumulative effect over distance, so it must be extremely small or else the universe would be ripped apart over large distances. Even at the vast scales of the Milky Way, the effect is unnoticeable. The repulsive effect only becomes noticeable at the scale of the whole universe.

So we need an astonishingly small value for the vacuum energy. However, when attempts are made to calculate the amount of vacuum energy, the result is very different to that required to be the dark energy. In fact, the value for the vacuum energy is calculated to be an enormous figure: a trillion trillion trillion trillion trillion trillion trillion trillion trillion trillion trillion trillion times larger than the observed value of dark energy density. And that's a lot of trillions! Famously, this has been called the "biggest error in the history of science". Vacuum energy is predicted to be far too large to be dark energy.

---

[9] For the technical details of this, see the superb paper by John C. Baez and Emory F. Bunn, *The Meaning of Einstein's Equation*, http://arxiv.org/abs/gr-qc/0103044

This sounds suspicious. The sums for this vacuum energy hypothesis simply do not add up. Not only do they not add up, they are astronomically wrong. It is possible that there is some mechanism which greatly reduces the vacuum energy, for example, the energy of different types of symmetrical particles might cancel each other out. But we would generally expect this cancelling mechanism to result in a zero vacuum energy, not a tiny value.

However, the problem with the necessary value of the vacuum energy is not that it is so much smaller than the predicted value. The problem is that it is set to a **particular** tiny value which results in an "interesting" universe. If the universe expanded at too fast a rate, galaxies would be unable to form. If the universe expanded too slowly, it would collapse back on itself before intelligent life could form. The amount of vacuum energy required is not just set to a much lower value than predicted, it is set to **just the right** much lower value than predicted. This gives an impression of fine-tuning of the vacuum energy.

There are more bizarre coincidences. The predicted density of dark energy in the universe is remarkably close to the density of matter in the universe. This is surprising as the density of matter reduces sharply as the size of the universe increases (the matter becomes more diffuse), whereas the dark energy density remains constant. The dark energy density remains constant purely because it is the energy of empty space: as the size of the universe increases, so does the amount of empty space. This apparent balance between the amount of mass and the amount of dark energy has been called the *cosmic coincidence problem*. This coincidence – this unlikely balance – appears to indicate that there is something very special about the current era.

In order to explain these apparent fine-tuning problems, the dreaded "multiverse" model has been introduced. A series of parallel universes are postulated with the vacuum energy being set to different values in each parallel universe.

However, most physicists would find such an idea highly fanciful. They may even accuse it of being unscientific.

There are many other problems with the theory of dark energy which I am not going to list here but I would direct you to the Wikipedia page on dark energy for a full list: "The identification of dark energy as a cosmological constant does not appear to be consistent with the data."

In short, the theory of dark matter appears convincing, but the theory of dark energy appears less than convincing. In the later chapters of this book, a simple alternative explanation for the acceleration of the expansion of the universe will be presented, an explanation which requires no fine-tuning of the vacuum energy to some bizarrely tiny value. I believe we are looking at the problem from the "wrong end". Instead of the setting of vacuum energy to a particular value which just coincidentally happens to produce an "interesting" universe, the hypothesis will be presented that such a universe is a fundamental necessity.

5

BLACK HOLES

Put simply, a black hole is an object of such extraordinary mass and gravitational pull that nothing can escape it – not even light. Just the name "black hole" conjures an image of a fearsome entity, dark and malevolent, stalking space ready to catch any unlucky passing object in its unbreakable grip. It seems like an object out of science fiction more than science fact. However, the existence of black holes is now a well-established reality.

Black holes are stars which have collapsed under their own weight. This can happen at the end of a star's lifetime. A star produces its energy via nuclear fusion, a process which fuses hydrogen nuclei together under immense temperature and pressure to become helium nuclei. The energy released is sufficient to overcome the immense gravitational force which continually tries to collapse the star. However, when a star burns off all its hydrogen fuel, it starts to collapse under its own weight. As a result of this extra pressure, the helium in the star starts to fuse. Energy is released, and the star expands in size to become a *red supergiant*. These are the largest stars in the universe, up to 1800 times the size of our Sun. Inside the red supergiant, progressively heavier and heavier elements are fused to produce energy. This process stops when iron is produced in the core. Iron is the equivalent of ash for nuclear fusion as it requires more energy to fuse iron than the process produces.

At this point, the red supergiant collapses and the outer layers of the star get blown off in a huge supernova explosion. A supernova is so bright that for a few days it can outshine all the other stars in a galaxy put together.

We considered how white dwarf stars explode as Type 1a supernovas in the previous section, describing how those supernovas make excellent standard candles. However, Type 1a supernovas never result in black holes – the exploding white dwarfs leave no remnant. The type of supernova we have described here, which start from red giants, are called core-collapse supernovas, otherwise known as Type II supernovas. These supernovas can leave behind a remnant at their centre, so only Type II supernovas can generate black holes.

What happens to the collapsing star now depends on how heavy the remnant is. For remnants of mass 1.5 to 3 times the mass of the Sun, the result of gravitational collapse after a supernova explosion can be a *neutron star*. Neutron stars are incredibly dense: six million tons to the cubic inch.

For even larger stars, if the mass of the remnant exceeds about three solar masses then the gravitational collapse can be unstoppable. In less than a second, the core of the star implodes from a ball about the size of the Earth to a ball about the size of a small city. And this implosion continues down to a small point. It is this point which becomes a black hole.

The black hole then increases in mass by attracting other mass from its surroundings. In this way, black holes can grow to possess extraordinary mass, and an extreme gravitational field to match.

BLACK HOLES

The wonderfully evocative name "black hole" was coined by John Wheeler in 1967. However, surprisingly, we can trace the first proposals back to the eighteenth century.

The idea was suggested independently by the British geologist John Michell, and the French mathematician Pierre-Simon Laplace. By using Newtonian physics they considered the escape velocity of an object trying to break free of a gravitational field around a star or planet.

This means an object is propelled off the surface of the planet with a velocity, v, and that initial kinetic energy must be greater than (or at least equal to) the gravitational energy pulling it back to the planet.

In which case:

$$\frac{1}{2}mv^2 = \frac{GMm}{r_{esc}}$$

where m is the mass of the object trying to escape, M is the mass of the planet, and r_{esc} is the escape velocity (the left-hand side of the equation is the kinetic energy, the right hand side of the equation is the gravitational energy). On rearranging the equation we find:

$$r_{esc} = \frac{2GM}{v^2}$$

For black holes, we put v equal to the speed of light, c, to indicate that any object trying to escape the clutches of a black hole would need an escape velocity faster than the speed of light (which is impossible, of course):

$$r_{esc} = \frac{2GM}{c^2}$$

This black hole radius calculated by the eighteenth century physicists remains an accurate formula still in use to this day.

Of course, light is not slowed down by gravity – light always travels at a constant speed. So light is actually trapped in a black hole by the extreme curvature of space.

This result of Michell and Laplace was ignored throughout the nineteenth century as it was not believed that light (which was known to be massless) was affected by gravity. It took the arrival of general relativity before the idea was revisited.

As we saw in the earlier chapter on gravity, merely from a thought experiment involving a window in a spaceship, Einstein was able to predict that gravity bent the path of light. This was a revelation: light might be massless, but it was still affected by gravity (gravity being the curvature of space itself). This again raised the possibility of an object whose gravitational pull was so strong that not even light could escape. Hence, from a mere thought experiment, Einstein predicted the existence of black holes. The existence of black holes is now an accepted fact in astrophysics – more evidence of the validity of general relativity.

Einstein's equations of general relativity published in 1915 were complex, non-linear, and very difficult to solve. In this respect, solving the equations is similar to solving non-linear equations of fluid dynamics, or weather prediction. Even to this day, exact solutions for Einstein's equations have only been found for very simplified models, with the solutions to more complicated situations only capable of being solved by number-crunching on extremely powerful computers. For example, it is now possible to simulate the results of black hole collisions using the Einstein equations and super-computers capable of performing one thousand billion calculations per second.

The very first exact solution to the equations was presented by Karl Schwarzschild just months after Einstein published his result. Schwarzschild used a simplified version of the Einstein equations to produce a *vacuum solution*, a solution for an empty space. This considered the curvature of empty space around a spherical mass, for example, a star. Further simplifications meant the star should not be rotating, and had no overall electric charge. These are considerable simplifications, but they did allow the equations to be solved.

The Schwarzschild solution predicted a black hole radius which is called the *Schwarzschild radius*. The value for this Schwarzschild radius is the same as that produced by the classical derivation for the black hole radius presented earlier:

$$r_s = \frac{2GM}{c^2}$$

Remember this formula. We will be seeing it again later.

Rather incredibly, Schwarzschild derived this solution while serving in the trenches for Germany in the First World War. Sadly, he was killed on the Russian front just months after publishing his result.

The Schwarzschild radius around the black hole is usually referred to as the *event horizon*. Once any object passes the event horizon into the black hole, it has passed the point of no return. It can never get out again.

If you were approaching a black hole in a spaceship, you might not even notice the moment you passed the event horizon. However, you have already signed your death warrant once you pass the event horizon. It is like falling over the edge of a waterfall on a river. There is no way you can swim back up. From that point on, you are doomed to be dashed on the rocks below.

As you pass further into the interior of the black hole, it is believed that the gravitational stresses on your body would increase. Your feet would be pulled into the centre faster than your head – a process given the wonderfully descriptive term *spaghettification*.

Although we can have no direct, certain knowledge of the behaviour of physics inside a black hole (we are fundamentally prohibited from obtaining information from inside a black hole), conventional theory suggests that material entering a black hole will continue to fall to the centre, which has become a region of unimaginable density. According to general relativity, at the very centre we will find that the density increases to an infinite amount, a point called a *singularity*. Current theories of quantum gravity say that the singularity would not occur in reality – it is only the shortcomings of our current theories which suggests singularities. This is because singularities are so distasteful to physicists: they are the point at which the laws of physics break down. When we find infinities emerging in theories it usually means that the theory is inaccurate, or has been pushed too far. Einstein never accepted black holes because he did not believe a singularity could occur. If we had the right theory, the singularity would surely disappear from our theory.

Bear this in mind, because we will be returning to this important point (and proposing a singularity-free theory) later.

Supermassive black holes

A black hole increases its mass by attracting other mass which surrounds it. In this way, the mass of a black hole can rise to an extraordinary value: as much as 450 times the mass of the Sun. In order to study these so-called "supermassive" black holes, we obtain our most accurate data from space telescopes which are satellites in Earth orbit. The most recently-launched space telescope is NuSTAR, launched in 2012.

Once in orbit, the NuSTAR telescope expands to a 10-metre focal length. The optics (lenses) are at one end of the mast, and the electronic detectors which capture the images are at the other end. NuSTAR is designed to capture images of X-ray emissions from distant high-energy sources such as exploded stars, clusters of galaxies, and black holes. NuSTAR is the first space telescope with the ability to focus X-rays to form sharp images. It is very difficult to reflect

X-rays as they tend to pass straight through any object they hit head-on, or get absorbed by the object (this is what makes medical X-rays possible). The mirrors of NuSTAR overcome this problem by just glancing the X-rays to modify their direction slightly. Because of this slight deflection, focusing the X-rays requires a distance of 10 metres between the mirrors and the detector. This is achieved on NuSTAR by using an extendable mast.

X-ray imaging (really just a high-energy form of light) is the best way to capture images of black holes. While black holes do not themselves emit X-rays, material falling into a black hole gets heated by the extreme gravitational field and X-rays are produced. This material, which is external to the event horizon, forms an *accretion disk* of debris which orbits the black hole at great speed (the rings of Saturn are another example of an accretion disk).

As Eliot Quataert of Berkeley's Theoretical Astrophysics Center explains: "That picture that matter gets sucked into a black hole, that's one of the biggest confusions that's out there, partially because of science fiction like *Star Trek*, and things like that. For matter that's far away from a black hole it actually doesn't get sucked in: it's very much like the planets in the solar system going around the Sun. Things just go around and around and around and around."

About 10% of these accretion disks can generate huge jets of matter which emerge from the polar regions of the black hole travelling at near to the speed of light (the mechanism by which this occurs is not yet understood). The radiation from these jets can be a thousand times more powerful than the luminosity of our entire galaxy, and these incredibly bright objects – the brightest objects in the entire universe – are called *quasars*.

BLACK HOLES

It is ironic that the brightest objects in the universe are actually black holes!

There is believed to be a supermassive black hole at the centre of our galaxy called Sagittarius A*. This has a mass of about 4.3 million solar masses. Everything in our galaxy is spinning around this supermassive black hole at its centre. It is now believed that there is a supermassive black hole at the centre of every galaxy, its huge gravitational pull essentially holding the galaxy together. So while black holes have a reputation as fearsome objects of destruction, they actually play a vital role in shaping the universe and holding it all together.

The more we learn about black holes, the greater the role they seem to play in the formation and structure of the universe. And this is a theme which we will find continued in this book.

Black hole thermodynamics

You might be surprised to learn that a science developed to analyse 19th century steam engines is now our main tool in analysing black holes. The science of thermodynamics was established as a means for studying the heat and motion of steam engines. Thermodynamics considers the energy and motion of atoms, and the heat and pressure that motion produces. We are now finding that the fundamental principles behind thermodynamics have particular relevance to the mechanisms associated with black holes. As Jim Al-Khalili says in his excellent BBC documentary *Order and Disorder: Episode 1* (available on YouTube): "As unlikely as it sounds, steam engines held within them the secrets of the cosmos."

In the 19th century, Britain's Victorian industry led the world. That industry was based on the newly-invented steam engine. The expansion of the British Empire around the world was powered by the steam engine, allowing the construction of the railways and steam ships, and providing a huge boost to manufacturing. Britain emerged as the world's most powerful trading nation.

In the early 19th century, Britain's defeat of Napoleonic France motivated humiliated French scientists to learn more about steam power in an attempt to catch up with the British technological lead. The French engineer Sadi Carnot is known as the "Father of Thermodynamics". Carnot was determined to analyse how steam engines work. In 1824, he wrote the legendary *Reflections on the Motive Power of Fire*. Carnot realised that heat would flow from a hot source to cooler surroundings, and that flow could be harnessed to do work.

But it was only when the principle of atoms was proposed that we could really make sense of heat and motion. It was realised that heat was nothing more than the random motion of atoms. If the atoms move more vigorously, then the object heats up. This provides us with a greater understanding of how heat moves. If one corner of an object is hot, then that heat will spread through the object so that the entire object becomes one temperature. This is due to the motion and collision of atoms within the object, but there is a deeper principle at work here to explain why the flow of heat works in only one direction.

If we have a small amount of gas molecules in one corner of a room, then, over time, that gas will spread over the entire space of the room. What is happening is that a very ordered state is changing to a very disordered state. When all the gas molecules were carefully positioned in the corner of the room, this represents an ordered state. Over time, when the room becomes full of randomly-moving molecules of gas, this represents total disorder.

This principle – that the state of a system will tend to move from order to disorder – is known as the *second law of thermodynamics*, and is one of the most important and fundamental principles in physics. It applies to all systems throughout the universe, from steam engines to black holes.

The second law of thermodynamics introduced a quantity of a system called its *entropy*. Entropy can be considered to be the amount of disorder in a system. The second law says that the entropy of a system only ever increase – it can never decrease. Essentially, the disorder of a system can only ever increase over time.

This principle – that disorder can only ever increase with time – seems so fundamental, so obviously correct, that the second law of thermodynamics is considered to be a completely unbreakable fundamental law of physics.

This one-way principle of entropy increase might remind you of a one-way property of a black hole. Can you think of anything about a black hole which always increases and can never decrease? Well, the mass of the black hole can only ever increase (nothing can ever leave a black hole). Which means the corresponding Schwarzschild radius – and, specifically, the associated area of the event horizon – can also only ever increase. Because of this similarity, in 1972, Jacob Bekenstein proposed that a black hole has entropy proportional to the area of its event horizon.

It might seem strange that we consider a black hole to have entropy at all. But, remember the second law of thermodynamics which says that entropy can only ever increase. It might appear that we can eliminate the entropy of matter by sending it into a black hole, in which case its entropy would appear to have gone for good. This sounds like a method for reducing entropy, which would break the unbreakable second law. So, to prevent the second law from being broken, the entropy of any mass entering the black hole gets added to the total entropy of the black hole. This is reflected in the increased area of the event horizon. In this way, entropy only ever increases.

Consideration of black hole entropy is perhaps the main tool we possess for analysing the properties and behaviour of a black hole. This area of study is called *black hole thermodynamics*.

Information and entropy

There is an important connection between the information contained in a system and the entropy of that system. A simple way to understand this is to imagine you are in a library, and there are ten different books on the table in front of you. Each book is about a different subject, and each book contains useful information. So each book can be clearly distinguished from any of the other books by means of the distinct useful information it contains.

Over time, however, the books start to decay. First the pages fade, then the paper rots. Eventually the whole book crumbles into a pile of dust. The entropy of each book has therefore massively increased. In fact, the entropy of each book is now at a maximum – they cannot get more disordered. All the useful information contained within each book has now been lost. The books can no longer be distinguished – all we are left with is ten piles of crumbly dust.

So as entropy increases, information is lost. If entropy is at a maximum, information is at a minimum, and if entropy is at a minimum then information is at a maximum. You might often read that "entropy is information" in the scientific literature. As we can see from our discussion, this is worse than misleading: it is simply wrong. Entropy is actually the **inverse** of information: entropy is a measure of information **loss**. We can consider entropy to be the amount of invisible information **lost or hidden from us**. As Seth Lloyd says in his book *Programming The Universe*: "Entropy, which is just invisible information, is also a measure of ignorance."

If we have a room full of gas, and we leave it for a while, the gas will all become the same temperature. At this point, the entropy of the gas is at a maximum. The molecules of the gas are flying around in a completely random manner. We cannot keep track of the individual gas molecules, so we can only describe the gas by a single value: its temperature. That single value is the only information we have left. At this point, we say the gas is in a random, *thermal* state, a state of maximum entropy.

Returning to consider our ten books, our definition of the amount of information that a book contains is rather dubious. We are saying that all the information in a book is contained in the letters and words of that book. We are, of course, unable to keep track of the position and motion of all the atoms in the books – that information is hidden from us. But Nature does not recognise letters and words. All Nature sees are the atoms and particles in the book. As far as Nature is concerned, the fundamental information contained within the book is the position and velocities of the atoms which make up the book. And that amount of fundamental information is, of course, retained even when the books crumble into dust. The book still contains the same number of atoms. Those atoms still have the same number of velocities. Even when we dig down to the quantum level, the mathematics of quantum mechanics insists that all the information about elementary particles is conserved (this is called *unitarity*). So this introduces a very important principle: at the fundamental level, **information is always conserved**.

The holographic bound

Any information (e.g., contained in a book) entering a black hole is clearly going to become hidden information – we are never going to be able to read that book again. So that hidden information acts to increase the entropy of the black hole. This means a black hole can really be considered to be a store of hidden information.

As we saw in the previous section, Jacob Bekenstein proposed that the area of the black hole's event horizon was proportional to the entropy of the black hole. This means that the area of an event horizon is proportional to the information hidden inside the black hole. This is a surprising result: the information is proportional to area. More likely we would have expected information to be proportional to the volume of the black hole, after all, the amount of information stored in a book is proportional to the number of pages – not the area of its cover.

What is more, Bekenstein was able to extend his result to calculate the maximum information content of **any** region of space. His logic was ingenious:

- Consider some matter (and its associated information) enclosed in a sphere.

- Some mass is added to force the region to collapse to a black hole.

- The entropy of the resultant black hole is proportional to the area of its event horizon. This area is obviously smaller than the area of the original region.

- Entropy never reduces, so the entropy of the original region must have been less than the entropy of the black hole.

- So the maximum information content of any region of space is less than a black hole of equivalent radius.

- Therefore, the maximum information content of any region of space is proportional to the area of that region.

This result – that the maximum information content that can be held in a region of space is proportional to the area of that region – might appear surprising. Intuitively, we might consider the maximum information content to be proportional to the volume of that region. This result was realised by Jacob Bekenstein, and this limit on information content is called the *holographic bound*. In his book *The Black Hole War*, Leonard Susskind says this result is "profound and probably holds the key to the puzzle of quantum gravity."

I am sure you have encountered holograms before. There are holograms on most driving licenses and credit cards.

They are the small shiny foil squares which seem to contain an image. Holograms are generated by splitting a laser into two beams: one beam illuminates the object, while the other beam shines directly on a photographic plate. As a result of the two beams, an interference waveform pattern is generated on the photographic plate. Inevitably, the pattern saved on the photographic plate appears completely scrambled, but when the photographic plate is illuminated again, the three-dimensional representation of the object is regenerated. So holograms reveal the general principle that a three-dimensional object can be accurately captured on a two-dimensional surface. This is what the holographic bound suggests about any region of space.

String theorists have latched onto the implications of the holographic bound and developed a theory which suggests that all the information in the universe is "painted" on the outer boundary of the universe (taking their lead from the assertion of the holographic bound that the maximum information content of the universe would be proportional to the area of its boundary). The theory suggests that you are a three-dimensional projection of a two-dimensional structure on the boundary of the universe. This is called the *holographic principle*.

However, Jacob Bekenstein is also sceptical of extending his principle to the entire universe: "My result is not applicable to horizons which are not event horizons." According to Bekenstein, the holographic principle could only apply if the universe itself was a black hole. And that surely could not be the case!

Or could it? We shall see in the later chapters. The answer might surprise you.

The black hole information loss paradox

In 1972, when Bekenstein proposed that black holes had entropy, no one really knew what to make of the result. That all changed in 1974 when Stephen Hawking made his greatest discovery. Hawking realised that black holes not only have entropy, they also have an associated temperature. This might come as something of a surprise. It might be imagined that a black hole is completely cold and dark. If a black hole has a temperature, then that means it emits radiation. How on earth could that occur?

So far in our analysis of black holes, we have only considered the effects of general relativity. In order to understand black hole radiation, we need to also consider the effects of quantum mechanics.

If you remember from the discussion about the vacuum energy in the last chapter, quantum mechanics allows enough energy to be borrowed for a short period of time to create particles in empty space. These virtual particles have to be produced in pairs: a particle and its antiparticle. For example, an electron and a positron may be produced – but only for a very short period of time. Very shortly after the pair of particles appear, they have to annihilate each other.

Stephen Hawking realised that if a pair of virtual particles was produced at the event horizon of a black hole, one of the particles might be sucked into the black hole while the other particle remained outside. In this case, there would be no way that the particle inside the horizon could re-emerge. The particles would be incapable or annihilating each other. The particle which remained outside the horizon would be free to be emitted as *Hawking radiation*.

The diagram below shows some virtual particles (on the left) being created and annihilating each other. However, on the right-hand side we see one of the virtual particles being sucked into the black hole. The remaining particle is emitted as Hawking radiation:

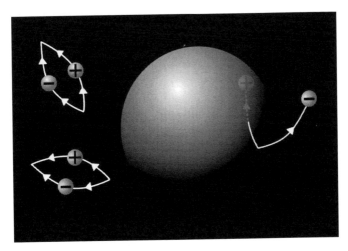

As surprising as Hawking radiation may appear to be, it is even more surprising to realise that this will inevitably result in the evaporation of the black hole. If a black hole is radiating energy, then its mass will inevitably reduce (through $E = mc^2$). Eventually, the black hole evaporates to nothing.

The evaporation process is incredibly slow and weak, because the predicted temperature of a black hole is only a fraction of a degree above absolute zero. It would take billions of years before any black hole would disappear through evaporation. Nevertheless, the fact remains that every black hole will eventually evaporate.

When he announced his discovery of black hole radiation, Stephen Hawking dropped another bombshell. He announced that when a black hole finally evaporates, all the information that ever entered the black hole would have vanished as well. This is because the radiation was considered to be completely thermal – random – and could not be responsible for carrying any information away from the black hole.

This was such a bombshell because, as we considered earlier, quantum mechanics states that fundamental information is always conserved. This principle is a cornerstone of physics. If information could genuinely be lost from the universe then this would represent a crisis in physics, and would cast doubt on our understanding of the basic principles of Nature. In essence, we have a clash between quantum mechanics which says that information must be conserved, and general relativity which says that nothing can come out of a black hole. So, we have our two great theoretical pillars which support the whole of physics, and, incredibly, one of them must be wrong – but we don't know which theory is wrong!

So physicists eagerly got to work trying to find a way out of what is now known as the *black hole information loss paradox*. The proposed solutions include:

- The radiation somehow manages to carry the information away from the black hole. However, the radiation has always been assumed to be thermal, i.e., completely random. There is no way for anything to leave the interior of a black hole.

- There is a microscopic remnant left behind after evaporation which includes all the information which has ever entered the black hole. This is generally considered impractical.

- The singularity forms a connection with a baby universe so that even when the black hole evaporates, all the information is maintained in the baby universe. Which is fine, but it effectively represents a loss of information in this universe.

None of these solutions is entirely convincing. However, one proposal has found widest acceptance, and we will consider it next.

Black hole complementarity

The principle of black hole complementarity was proposed in 1993 by Leonard Susskind, Larus Thorlacius, and Gerard 't Hooft as a solution to the black hole information loss paradox. Complementarity avoids the paradox by suggesting that the information both enters the black hole and is **also** reflected at the event horizon. This appears to imply there are two copies of the information. How can this be true?

Well, it all depends on your point of view. When an observer falls through a black hole event horizon, he might not feel anything out of the ordinary to tell him he is irretrievably doomed. As he is in gravitational free-fall, Einstein's equivalence principle means he does not even feel the pull of gravity. However, for an observer outside the event horizon, the situation is different. To an observer outside the event horizon, the infalling object appears to slow down and stop at the horizon. Time itself appears to stop for the infalling object. This is because it becomes progressively more difficult for light to escape from the black hole. When the object reaches the event horizon, light can no longer escape and the object appears to freeze in time at the horizon.

So we have two different views of reality depending on where the observer is positioned. For an observer outside the event horizon, the infalling object appears to freeze at the horizon, but the experience of the infalling object is different as it notices no change as it passes the event horizon and continues unhindered toward its doom at the singularity.

The diagram below shows these two points of view. Observer A is outside the horizon, and sees the infalling object (usually chosen to be an elephant for some reason) freeze at the horizon. It is then broken down due to the temperature of the black hole and is radiated away into space. So the information is preserved in the radiation. Observer B is inside the event horizon and sees the object continue unhindered towards the singularity:

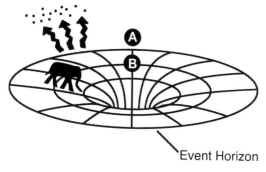
Event Horizon

However, there is a problem with this Xerox machine at the event horizon. Quantum mechanics prohibits the precise copying of any object. If we could take a precise copy of a particle, for example, we could take one particle and measure its velocity, and we could take the copied particle and measure its momentum. Hence, we could know both the velocity and momentum of the particle – something prohibited by the Heisenberg uncertainty principle. This law prohibiting precise copying is called the quantum *no-cloning theorem*.

Black hole complementarity has a clever way of avoiding the no-cloning theorem. Each observer – observer A outside the horizon, and observer B inside the horizon – can only see one copy of the information. Observer A can see the information in the radiation, while observer B can see the information in the unharmed elephant. But, crucially, neither observer can see both copies of the information.

So black hole complementarity seemed like a very satisfactory solution to the information loss paradox. In 1997, Stephen Hawking had made a public bet with John Preskill that information was definitely lost in black holes. However, in 2004 Hawking announced that he was conceding the bet – so even Stephen Hawking was convinced that the information came out with the radiation. For winning the bet, John Preskill received a baseball encyclopedia from Stephen Hawking (because it was heavy and difficult to get information out of it).

However, it now appears that Hawking conceded the bet prematurely. In the summer of 2012, the physics community got very excited when a group of physicists based in California (Almheiri, Marolf, Polchinski, and Sully – collectively known as AMPS) realised there was a flaw in complementarity.

According to AMPS, the observer outside the event horizon could obtain his information about the infalling object from the emitted radiation. But the flaw in complementarity was based on the fact that the observer could then jump into the hole and catch up with the object itself. The observer would then have two copies of the information, and that breaks the no-cloning theorem.

The only way out of this paradox is, according to AMPS, that any object entering the black hole is no longer unaware of the moment it crosses the event horizon. The moment the object crosses the event horizon there has to be a dramatic instant incineration of the object by Hawking radiation. This appears to be the only way to ensure that the no-cloning theorem is not violated. The AMPS theory has obtained the nickname "firewalls" because of this.

But why should a firewall appear out of empty space? No one seems happy with this conclusion, not even AMPS. There have been a host of different papers published over the last few months, all arguing different viewpoints. There is no consensus of opinion. In fact, this area of research seems to be in a state of chaotic flux. Most of the papers seem to be tinkering around the edge of the event horizon, trying to save the old model. The information paradox seems to be as far away from being solved as it has ever been.

I think we can safely say the information loss paradox remains unresolved. As I said earlier, either quantum mechanics or general relativity must be wrong. I believe it is general relativity which requires modification. Later in this book we will be revisiting the paradox armed with a new theory, a theory which does away with the tinkering and proposes a radical reassessment of general relativity and our model of black holes.

6

THE SOAP BUBBLE UNIVERSE

So far in this book, the material has not deviated from orthodox physics and cosmology. At this point, I have to warn you that we are now going to take a different turn. Over the next few chapters, a startlingly original hypothesis will be presented. We will build toward the hypothesis in a logical manner, and I firmly believe the hypothesis has a firm grounding in orthodox physics. However, the nature of the hypothesis is inevitably far removed from the mainstream. If the hypothesis appears too far-out for your liking, I can only ask you to bear with me as we explore the remarkable implications and apparent success of the hypothesis. Clear evidence will be presented. I hope you will be won over.

Sometimes you can stare at a problem for too long that you become unable to see a solution. Sometimes you need a radically different way of looking at a problem.

Up to now, we have studied several of the greatest mysteries in physics. We have considered the mystery of gravity, the puzzling acceleration of the expansion of the universe, problems with the inflation hypothesis as a means of explaining the flat universe, undesirable singularities in black holes, the mystery of black hole entropy being proportional to the surface area of the event horizon, and the apparent loss of information in black holes. It is now time to present a single, simple, logical solution which we shall see has the potential to solve **all** these problems.

There is a remarkable result in physics that has been known for many decades. The result seems to point to something very important and profound about the universe, yet it appears that no one quite understands or knows what to do with the result. It is the principle that the universe has zero total energy.

But how on earth can the universe have zero total energy? The universe obviously contains mass, and we know through $E = mc^2$ that mass can be converted to energy. So how can the total energy in the universe be zero? In order to understand this, we have to think a little bit more about how the force of gravity works.

Consider two masses which are quite close together. If we try to separate those two masses, we will have to perform work against the force of gravity. In this respect, the force of gravity between the two masses functions very much like a spring, the spring (gravity) attempting to pull the two masses back together:

When you move the two masses apart, energy is stored in the gravitational field: energy is stored in the gravity "spring". The density of the energy is actually proportional to the square of the field. How can energy be stored in a spring? Well, just imagine a traditional clockwork wristwatch. When you wind the wristwatch, you transfer energy to the spring, and it is the spring which stores the energy. Slowly over time, the energy in the spring is released and can power the motion of the hands of the wristwatch. And it appears to be a very similar situation with gravity. When masses are pulled apart, energy is transferred to the "spring" of the gravitational field. Energy is stored in the field, and then over time that energy is released to generate motion, i.e., to bring the masses back together again.

But, when you think about it, you realise this "spring" of gravity does not behave like a normal spring. With a normal spring, the further apart you pull the two objects, the greater the force the spring exerts to pull the objects back together. So, with a normal spring, it becomes ever harder to pull the objects further apart. Gravity does not work like that. If you consider the equation for Newton's law of gravitation, you will see that the force between the masses reduces as the distance between the masses increases. This behaviour is therefore different to the behaviour of a mechanical spring, and it is very hard to imagine a spring behaving like that. The spring would have to be very hard to pull apart initially, but it would get progressively easier as you pull the masses further apart. In this respect, the "gravity spring" is a very peculiar type of spring.

But it is this peculiar behaviour of the gravity spring which enables the universe to have zero total energy. This is possible because the behaviour of the gravity spring means the gravitational energy contained within the spring (i.e., the energy contained within the gravitational field) can be considered to be negative energy.

So how can gravitational energy be negative? Well, if objects are separated to infinity they feel no gravitational pull between themselves, so the gravitational energy of the system is zero in that case. But when those objects were initially clumped together, you had to put a lot of energy into the system to force them apart (remember: the gravity spring does not work like a normal spring – it is very hard to pull objects apart initially). So if you have to put energy into a system just to get to a zero energy situation, this means the energy of the system when those objects were initially clumped together must have been negative.

So gravitational energy is negative. But what is even more remarkable is that when we consider the actual data for the energy of the universe, the positive energy of the universe is precisely matched by the negative gravitational energy – giving a total energy for the universe of zero. As Richard Feynman said in one of his lectures on gravitation in the 1960s:

> *If now we compare this number (total gravitational energy M^2G/R) to the total rest energy of the universe, Mc^2, lo and behold, we get the amazing result that $M^2G/R = Mc^2$ so that the total energy of the universe is zero. Why this should be so is one of the great mysteries – and therefore one of the most important questions in physics. After all, what would be the use of studying physics if the mysteries were not the most important things to investigate?*

So Richard Feynman considered the mystery behind the zero-energy universe to be "one of the most important questions in physics". Trusting his instincts, he knew this undoubtedly held the key to a principle of the greatest importance.

Perhaps there is a simple and profound way to realise that the total energy of the universe must be zero. And this method can be derived from our fundamental principle that there is "nothing outside the universe". This principle is beautifully expressed by Misner, Thorne, and Wheeler in their classic textbook *Gravitation*:

THE SOAP BUBBLE UNIVERSE

There is no such thing as the energy (or angular momentum, or charge) of a closed universe, according to general relativity, and this is for a simple reason. To weigh something one needs a platform on which to stand to do the weighing.

There can be no such weighing platform outside the universe – because there is nothing outside the universe. You could never put the universe on weighing scales to determine the total mass-energy of the universe:

Those who have suggested that there is a non-zero value for the total energy of the universe are surely committing a *category error*: taking laws which only apply to objects within the universe and applying those laws to the entire universe itself. These are entirely different circumstances, and cannot be compared.

What would a numerical value for the total energy of the universe actually mean? How could the energy be used?

Obviously it could not be used to power anything outside the universe!

There is a also a famous equation derived from general relativity called the Wheeler-DeWitt equation which suggests that the total energy of the universe is zero.

Another convincing argument that the energy of the universe is zero comes from the law of conservation of energy. By having a zero-energy universe, we find that energy is conserved over the period before and after the Big Bang. This might seem a very strange thing to say, after all, it would appear there was nothing at all in existence "before" the Big Bang, so how could energy possibly be conserved in comparison with the era after the Big Bang? Well, as Alan Guth explains in his book *The Inflationary Universe*: "If the creation of the universe is to be described by physical laws that embody the conservation of energy, then the universe must have the same energy as whatever it was created from. If the universe was created from nothing, then the total energy must be zero."

Basically, what this means is that the universe must have had zero energy before the Big Bang (as absolutely nothing was in existence), so the universe must also have zero total energy in the era after the Big Bang. This might sound like quite a trivial argument, and there is a fair amount of controversy about whether energy is necessarily conserved in general relativity, but I believe the argument is actually extremely convincing.

We can surely proceed with a fair degree of confidence that the total energy of the universe is zero. And because we have derived this result from our fundamental principle, we can be sure that this result must apply in all conceivable universes.

THE SOAP BUBBLE UNIVERSE

Now let's try and generate a few equations to describe this principle that the total energy of the universe is zero.

If we can consider gravitational energy to be negative, and we know from our fundamental principle that the total energy of the universe is zero, we can equate the total mass-energy of the universe (obtained from $E = mc^2$) with the total gravitational energy:

$$M_U c^2 = \frac{GM_U^2}{R_U}$$

where R_U is the radius of the universe, M_U is the mass of the universe, c is the speed of light, and G is the gravitational constant.

You might recognise this equation as the equation referred to in the earlier quote from Richard Feynman.

Reorganising the terms in this equation leads us to a formula which appears to provide a means for determining the radius of the universe:

$$R_U = \frac{GM_U}{c^2}$$

This equation seems to be a quite remarkable formula: the ability to determine the radius of the universe from a combination of constants. It is as if the universe necessarily has to expand to this radius.

As I stated earlier, no one seems quite sure what to make of this result that the total energy of the universe appears to be zero. However, I believe this equation represents the first (but not the last) prediction of the zero-energy universe: it predicts the natural value of the eventual radius of the universe.

(The more foresighted reader is probably already realising that this necessitates a radical modification to the way the universe is predicted to expand – but more of that later.)

Note that, at this stage of the book, no particular force is being proposed as the reason for this expansion of the universe. That will come later. All that is being stated is that the universe necessarily has to expand to this radius: it is a logical necessity. This reflects a very important personal conviction I have. It is the **principle** which is fundamental (in this case, the principle of the zero-energy universe). The force (i.e., the movement of objects) arises necessarily in order to satisfy the fundamental principle. So the force as such is not really fundamental – it is secondary. It is the principle which is fundamental. The fundamental principle is a statement which is obviously correct. The fundamental principle **must** be true.

According to the dark energy hypothesis, the vacuum energy will continue the expansion, and the expansion can potentially continue forever. However, if we allow the universe to continue to expand due to the presumed influence of the vacuum energy then, as the radius progressively increases, we will inevitably break Feynman's wonderfully mysterious equation for the zero-energy universe. And we are not going to allow that to happen – it is a fundamental principle.

So, for the purposes of this analysis, we are going to forget about the vacuum energy hypothesis for the expansion of the universe. We discussed the problems with the troublesome fine-tuned vacuum energy back in Chapter Four. The value of the vacuum energy would have to be an extraordinarily small fine-tuned value, and when we try to calculate the value we just get a silly result which is far too large. In reality, the value for the vacuum energy could be arbitrarily close to zero. So we are entitled to consider alternative explanations for the expansion of the universe. We are going to consider the situation in which the

THE SOAP BUBBLE UNIVERSE

expansion of the universe is **not** due to the vacuum energy – instead it is due to the logical necessity that the total energy of the universe must be zero.

Returning to consider the previous formula for the necessary radius of the universe, the more eagle-eyed of you will recognise this resembles a formula which was introduced in the previous chapter: the formula for the Schwarzschild radius of a black hole:

$$r_s = \frac{2GM}{c^2}$$

What a surprise! Where did that come from? What has the radius of our universe got to do with the radius of a black hole? Well, actually, it is not so surprising to see the formula for the Schwarzschild radius appearing here. Remember how we derived the Schwarzschild radius by equating kinetic energy with gravitational energy? Well, we have arrived at this result in a similar manner by predicting that the total mass-energy of the universe should be equal to the gravitational energy, in order to produce a zero-energy universe. So we arrive at this equation when we consider the balance of energies. In the case of the black hole, the balance of energy is between an object trying to leave the hole, and the gravity which is pulling it back. Similarly, with the universe, we get a balance between the gravitational energy of the universe and the energy which is contained within the masses of the universe.

More and more we are discovering that the balance – and imbalance – of energies is the key to uncovering why objects move, and why the universe takes the form it does.

So a very nice analogy can be formed between the universe and a soap bubble! If you remember back to our analysis of a soap bubble in the introduction, we considered how the film of a soap bubble is elastic, and the potential energy of the surface is proportional to its area. Nature tries to minimise the potential energy, so the film tries to contract as far as possible, but this contraction is resisted by the outward pressure of the air within the bubble. Similarly for the universe, at the equilibrium distance we find a balance between the gravitational energy of the universe and its mass-energy. As the 19th century physicist Lord Kelvin said:

> *Blow a soap bubble and observe it. You may study it all your life and draw one lesson after another in physics from it.*

Returning to consider our formula for the radius of the universe, let us replace our Newtonian derivation with the more accurate Schwarzschild formula which we derived earlier:

$$R_U = \frac{2GM_U}{c^2}$$

Is this result matched by observation? Does it provide an accurate value for the radius of the universe? In order to check this equation, we substitute values for G, M_U, and c into the equation. We will take a value of 10^{54} kg for the mass of the universe.[10]

---

[10] The most common value for the mass of the universe is in the region of 3.3×10^{54} kg. However, this value includes the dark energy "fudge factor", as described back in Chapter Four. The dark energy value is included just to "balance the books" to produce a flat universe. We do not want to include this fudge factor in our calculations, especially as the theory introduced later in this book proposes an alternative to dark

THE SOAP BUBBLE UNIVERSE

This gives us a predicted value for the radius of the universe, R_U, of 1.5×10^{27}m which compares reasonably well with the actual measured value of approximately 4.3×10^{26}m. There is a discrepancy, but, as I said earlier, cosmology is not an exact science and it is hard to get accurate measurements. So we can say that there is a fair degree of quantitative evidence that the hypothesis is correct: the radius of the universe has to expand to its Schwarzschild radius. The equation seems to work.

Can this accurate prediction of the radius of the universe really just be a coincidence? Surely not. As Feynman suspected, this equation is telling us something profound.

In fact, the data seems to indicate that the radius of the universe is rather less than its Schwarzschild radius. In other words, this seems to imply that the universe fits the criteria for a black hole! So is the universe really a black hole? We will return to consider this later.

---

energy. So we will remove 70% of the mass of this dark energy universe to produce our value of 10^{54} kg.

The universe must be flat

So we have predicted that the radius of our universe should be equal to its Schwarzschild radius – and this compares with observation very well. But if the radius of the universe is equal to its Schwarzschild radius then we can make another prediction, and that is that the universe is spatially flat.

This is actually a well-established property of the Schwarzschild radius. If you remember in Chapter Five when we derived the Schwarzschild radius we considered the escape velocity of a particle. We found the escape velocity by equating kinetic energy with gravitational energy: a particle had enough energy – and only just enough energy – to leave the universe. This represents a flat universe. If the universe was closed, the particle would have looped back round into the universe again. If the universe was open, the particle would have left the universe at speed. So, again, it is all about the balance of energies.

Here is a bit of arithmetic which you should be able to follow.

Another way to show that the universe is flat at the Schwarzschild radius is to use the first Friedmann equation which we considered in Chapter Three. We saw there that the critical density of a flat universe (i.e., perfectly balanced between being open and closed) was given by:

$$\frac{3H^2}{8\pi G}$$

THE SOAP BUBBLE UNIVERSE

We also know from basic geometry that the volume of a sphere is given by:

$$\frac{4}{3}\pi r^3$$

so the volume of the spherical universe is:

$$\frac{4}{3}\pi R_H^3$$

where R_H is the Hubble radius, the radius of the observable universe.

So we know the density of a flat universe, and we know the volume of the universe. We can combine these two results to find the mass of a flat universe via mass = density × volume:

$$M_U = \frac{3H^2}{8\pi G} \times \frac{4\pi R_H^3}{3}$$

Substituting for $H = c/R_H$ (as was discussed in Chapter Three):

$$M_U = \frac{3c^2}{8\pi G R_H^2} \times \frac{4\pi R_H^3}{3}$$

which, if you reorganise the terms of this equation and cancel a few terms, gives:

$$R_H = \frac{2GM_U}{c^2}$$

which is, once again, the formula for the Schwarzschild radius. So a flat universe has a radius equal to its Schwarzschild radius.

So we no longer need inflation to solve the flatness problem (as discussed in Chapter Three). If you remember, the flatness problem arises because it appears that the density of mass and energy is precisely fine-tuned to result in a flat universe. The radius of the universe is treated as a "free parameter", which means it could take any possible value. This leaves us with the problem of why the universe is flat, and the inflation hypothesis emerges as a possible solution.

If, however, we consider there to be a logical necessity for the total energy of the universe to be zero, then the radius of the universe has to be equal to its Schwarzschild radius. This means the radius of the universe is no longer a free parameter – the universe must expand to a certain value, and that value ensures a flat universe will result. So, effectively, our fundamental principle bypasses the fine-tuning problem, and we no longer need to introduce inflation as the reason for the flatness of the universe.

Generally, the fact that the radius of the universe is very close to its Schwarzschild radius, and the fact that the universe is spatially flat, is presented as a lucky coincidence. However, we can see from our fundamental principle – which predicts a zero-energy universe – that a flat universe with a radius equal to the Schwarzschild radius is to be expected. It is no lucky coincidence. It is as though the universe must logically and necessarily expand toward the following scenario:

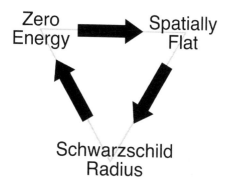

The hybrid expansion

Of course, if it is suggested that the universe will expand to a natural radius – like a soap bubble – then this clearly requires a radical modification to the existing model of cosmological expansion.

In the earliest stages of the universe, we could imagine the mass and energy of the universe being positioned in close proximity. Our analysis would suggest that, at this early stage, the total mass-energy of the early universe would be vastly greater than the gravitational energy contained in the field (gravitational energy is always negative – as we considered earlier – and at the time of the Big Bang with all matter so close together it would have been hugely negative):

This imbalance would result in a universe whose total energy would be far from zero. Nature could not tolerate such a thing – our fundamental principle says as much. There would be a logical necessity for the universe to rapidly expand until it reached the equilibrium point: the Schwarzschild radius. This initial acceleration is analogous to the situation believed to occur at the Big Bang.

Only after the universe has expanded to its Schwarzschild radius would the imbalance of energies be corrected:

THE SOAP BUBBLE UNIVERSE

So the hypothesis predicts a theory for the expansion of the universe which is a peculiar hybrid of the Big Bang theory and the Steady State theory. The expansion is predicted to accelerate from a point of great density until it reaches its Schwarzschild radius. Even when it reaches that point, the universe would continue to expand – probably for billions of years. However, once it passes the Schwarzschild radius it does start to decelerate. Eventually, the expansion would cease, and then reverse. The universe eventually settles at a constant radius. It really would be a "soap bubble" universe.

I do not believe anything similar to this hybrid expansion model has ever been suggested before. Undoubtedly, this is because it contradicts general relativity. As we saw in Chapter Three, general relativity either predicts a universe which collapses back in on itself or expands forever. The steady state position – with the universe remaining at a constant average radius – is unstable. However, the hypothesis presented in this book suggests that a stable equilibrium distance is not only possible, it is a certainty. We will consider the implications for general relativity in the next chapter.

It is time to present a new theory …

7

GRAVITY REVISITED

In the previous chapter, we saw how there could well be a logical necessity for the universe to have zero total energy. This could only be achieved with the assistance of gravity. Gravity was considered to provide a form of negative energy which could balance the positive mass-energy in the universe. The combination of the logical necessity for zero energy together with the theory of gravity meant the universe expanded to a certain radius.

The simplest way to think about this is to compare the universe to a soap bubble. The elastic film of the bubble always tends to contract (analogous to gravity), whereas the air pressure in the bubble is always directed outward (analogous to the mass-energy in the universe). The bubble eventually settles down to an equilibrium radius at which the two energies are balanced.

Considering the universe, it appears we could combine the logical principle and the conventional force of gravity together to produce a theory of *modified gravity*. This modified theory of gravity would act to move the radius of the universe to its equilibrium distance.

As we shall see, what makes this theory of modified gravity so compelling is not just that we have derived it in a logical manner, but because simple solutions to so many of the most intractable problems in cosmology fall naturally out of the theory. We have already seen how the theory predicts a flat universe, without fine-tuning or inflation. We will later see the remarkable implications for dark energy and black holes.

In order to consider the behaviour of this theory of modified gravity, we are going to use the spring analogy. It would appear the theory of modified gravity, like a normal spring, now appears to have a natural extension at an equilibrium distance:

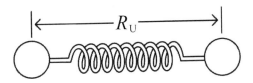

If a mass is moved, the spring is stretched. Energy is transferred to the gravity spring, so the balance of energy is upset. Gravity then works to restore the natural equilibrium distance. This effect reveals itself as an attractive force between the masses:

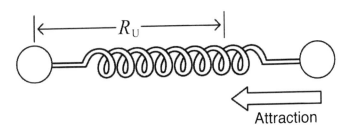

This would be, by far, the most common observed effect of gravity in the universe. For example, the Earth attracting the Moon, trying to drag the Moon back to the Schwarzschild radius of the Earth (which is a very small distance). However, by analogy with a normal spring, this "gravity spring" can also be compressed as well as stretched. If the mass is moved closer to the other mass so that it is inside the Schwarzschild radius, the balance of energy is disrupted once again, but this time in the opposite direction. Gravity again works to restore the natural equilibrium distance, but this time the effect would reveal itself as a **repulsive** force between the masses:

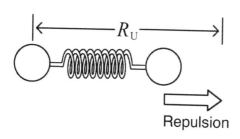

At this point, you might very well throw up your hands in horror at the prospect of modifying gravity. The theory of general relativity is regarded as one of the greatest achievements of the human intellect. It has been tested many times (remember Eddington's solar eclipse?) and always found to be perfectly accurate. So it might come as a surprise to find out that almost all physicists would agree that the theory of gravity has to be modified!

Gravity has to be modified – but only at the smallest scales. Though gravity is extraordinarily weak, at the smallest distances it is predicted to increase to a huge force. These distances are in the region of the Planck length, the smallest distance at which our laws of physics still hold. In order to accurately predict the behaviour of gravity at these distances, a new theory of quantum gravity will be required.

So gravity needs modifying at small scales, but at large scales it remains a fact that general relativity remains wonderfully accurate. This might be proving somewhat frustrating for a group of physicists who believe that the accelerating expansion of the universe might be explained by modifying gravity at large scales. If the large-scale theory of gravity could be modified to become repulsive at large distances then it could eliminate the need for the proposed dark energy. Sean Carroll expresses this in a video entitled *Dark Energy, or Worse: Was Einstein Wrong?* (available on YouTube):

> *We only ever infer the existence of dark matter and dark energy through gravity. We have been given a theory of gravity by Einstein – general relativity – which has been tested very precisely in the Solar System, but now we are applying it to regimes that are orders of magnitude away from that. How do we know that our theory of gravity works on those scales? Well, we don't. This is not a situation where professional cosmologists are trying to save Einstein's honour by saying that his theory must be right even though the data contradicts it.*

However, general relativity has proven just too accurate at large scales. These theories suggesting modifying gravity at large scales have not achieved much success or acceptance.

The novel feature of the theory of modified gravity being presented here is that for almost every object in the universe the theory does not suggest that the large-scale behaviour of general relativity is modified in any way. This is essential. The long-distance behaviour of general relativity is too well established. On the contrary, for almost all objects in the universe the theory of modified gravity presented here suggests a modification of gravity at **small** scales, the distances where everyone agrees that modification is essential. This is because the theory modifies gravity at distances inside the Schwarzschild radius of an object, and for most objects in the universe that is an extremely small scale indeed.

To put this into context, the Schwarzschild radius for the Earth is approximately 9mm. To see the repulsive gravity effect in reality, all the mass of the Earth would have to be compressed to the size of a peanut, and even then the effect would only be detectable inside the peanut. The Schwarzschild radius for the Sun is 3km. The Schwarzschild radius for a human being is less than the radius of an atom. So this repulsive gravity is not an effect which is going to be easily detectable. In fact, for just about every object in the universe, gravity would only ever manifest itself as an attractive force and standard general relativity would apply.

However, for the universe as a whole, the story is very different. The Schwarzschild radius of the universe is the huge distance of 1.5×10^{27}m (calculated from the mass of 10^{54}kg) – a distance which is larger than the radius of the current observable universe. So the whole mass of the universe is contained within its Schwarzschild radius. This means that, for the universe as a whole, the repulsive effect becomes highly-significant. Gravity would act to expand the universe.

We discussed the vast scale of the universe in Chapter Three, and here we see why this vast scale is so important. This ingenious hypothesis presented in this book suggests that gravity should be modified only at small scales. However, the universe is so vast, so utterly beyond our comprehension, that, for the universe as a whole, the small-scale becomes large-scale. So for the universe itself – and only the universe itself – gravity acts as a repulsive force.

The accelerating expansion (revisited)

As we discussed in Chapter Four, at the largest cosmological scale we find the accelerated expansion of the universe which is commonly associated with the presence of the mysterious dark energy. But the simple and elegant theory presented in this book suggests that this repulsive gravity effect is to be expected for any object whose mass is entirely contained within its own Schwarzschild radius – and that includes the universe. So the accelerated expansion of the universe could possibly be explained by this wonderfully simple principle of gravity acting as a repulsive force inside the Schwarzschild radius.

In which case, we are provided with a simple explanation of why galaxies are accelerating away from each other: they are simple "falling away". Just as when you drop a rock, it accelerates under the pull of gravity, so these galaxies are basically "falling" under the pull of gravity, and accelerating in just the same way as a dropped rock.

As another interesting thought, the Big Bang has always been presented as an initial explosion. After that initial impetus, gravity is predicted to take over and progressively slow the expansion of the universe, gravity being predicted to always act as an attractive force. So this explosive Big Bang set matter and energy on an inertial trajectory, acting very much like a ballistic bullet being shot from a gun.

However, this proposed new model of a hybrid expansion does away with the immense initial explosion and considers a universe with a slower initial acceleration due to gravity. Instead of the bullet being fired from a gun, imagine the bullet being dropped from your hand. The initial speed of the bullet will be low, but the bullet will accelerate under the influence of gravity and – given enough distance – the bullet will end up moving as fast as the bullet which was fired from the gun.

Similarly for the universe, the expansion would start slowly, and progressively pick up speed. This slower take-off provides an opportunity for the equalisation of temperature throughout the universe, resulting in the uniform distribution we see today.

So this theory avoids the need for inflation to explain the flat universe (the flatness problem) and avoids the need for inflation to explain the uniformity of the universe (the horizon problem). It appears there is no need for inflation.

So maybe the total universe is not a hundred sextillion times larger than the visible universe after all (as predicted by inflation).

If you remember back to our discussion of the dark energy hypothesis, we encountered the "cosmic coincidence problem" (just one of the fine-tuning coincidences associated with the dark energy hypothesis). This suggested that we are living in a very special era because the amount of mass and the amount of dark energy appear to be closely balanced at the moment (dark energy is predicted to carry on powering the expansion of the universe until it dominates the amount of mass). However, the hypothesis presented in this book predicts a universe which settles down to a constant steady state – in which case there would be nothing special about the current era. It is just one more coincidence eliminated by the hypothesis.

As well as considering the predicted effect on the whole universe, it is also important to consider the effect of the theory on the largest objects which exist within the universe: galactic clusters. Galactic clusters are groups of about 50 galaxies held together by gravity, and they represent the largest known gravitationally-bound objects. For the theory of modified gravity proposed in this book to be correct, it is essential that gravity is shown to be an attractive (not repulsive) force for galactic clusters. In this sense, the theory proposed in this book is very much a falsifiable theory: if the theory predicts that gravity should be a repulsive force for galactic clusters, the theory would have to be rejected (it is considered a good thing for a scientific theory to be falsifiable).

Fortunately, when we consider galactic clusters we find that the theory predicts gravity to be an attractive force. This is because the radius of a galactic cluster is in the order of 10^{22}m, and its Schwarzschild radius is much smaller at 10^{16}m. Hence, the theory has passed this test: galactic clusters are predicted to be held together by gravity.

All the galaxies within the cluster are predicted to attract each other according to conventional gravity. This is because the Schwarzschild radius of a galaxy is much smaller than the radius of the galaxy. The only objects which would exhibit gravitational repulsion would be those objects whose entire mass can fit inside their Schwarzschild radius.

The following graph shows the radius of various celestial bodies of different scales plotted against the Schwarzschild radius of those bodies. In this way, we can see if gravity is predicted to be an attractive or repulsive force for those bodies. The bodies all tend to lie very close to a straight line (shown dotted), as the bodies are all made of atoms and the line represents constant atomic density.

For most of the graph, the radius of the objects is greater than their Schwarzschild radius, and so gravity is an attractive force. However, in the shaded area, the radius dips to become less than the Schwarzschild radius, so gravity is predicted to be a repulsive force for these objects. Only the universe as a whole lies in this area, so only the entire universe experiences repulsive gravity:

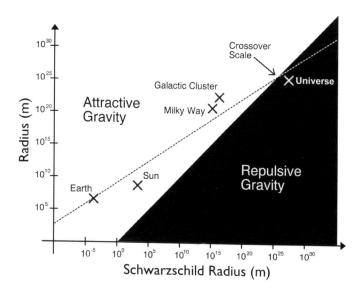

The point at which gravity switches from attractive to repulsive is called the *crossover scale*. For a successful theory which explains the acceleration of the expansion of the universe, it is generally accepted that the crossover scale should be in the region of the Hubble scale. From the previous graph, it appears the proposed crossover scale of approximately 10^{24}m would be absolutely perfect to explain the acceleration of the expansion.

This is by no means the only theory of modified gravity which attempts to explain the acceleration of the expansion by modifying general relativity. Let us consider what appears to be the leading modified gravity theory which is called the DGP theory, named after the surnames of the inventors, Dvali, Gabadadze, and Porrati.[11] It is interesting to compare the predictions of the hypothesis presented in this book with the predictions of the DGP theory.

The DGP model is based on string theory. According to string theory, particles are composed of one-dimensional strings. But it is also possible for there to exist objects with a greater number of spatial dimensions. For example, a two-dimensional surface is called a *brane* (from "membrane"). Branes can have any number of dimensions. In particular, the string theory model suggests that our four-dimensional universe might be a brane inside a higher-dimensional space called the *bulk*. Other parallel universes might be separate branes inside the bulk.

This model has some interesting things to say about gravity. It proposes that the reason that gravity is such a weak force is because much of its power leaks out of our universe into the bulk. This leakage is not supposed to affect the other three fundamental forces which explains why they are so much stronger than gravity.

The DGP theory of modified gravity attempts to explain the accelerated expansion of the universe by proposing that gravity is weaker at very large distances. The theory proposes that the universe's four spacetime dimensions are on a brane which is actually in a five-dimensional space. At shorter distances, the four-dimensional term dominates which gives rise to the conventional strength of gravity. However, at very large distances, gravity leaks into the fifth dimension, explaining why it is a weaker force at cosmic scales.

However the crossover point in DGP theory at which gravity behaves differently was calculated to be only the size

---

[11] G. Dvali, G. Gabadadze, M. Porrati, *4D Gravity on a Brane in 5D Minkowski Space*, http://arxiv.org/abs/hep-th/0005016

of the solar system, which is far too small. The DGP team suggest an ideal crossover scale for their theory would have been the vastly larger figure of 10^{21}m. As we have just seen, the crossover scale for the theory proposed in this book occurs at precisely the suggested scale. So the theory of modified gravity presented in this book more accurately models the universe than the most popular current theory.

So this is a really ingenious, simple theory, which matches available data. And this is not an *ad hoc* theory which has been invented to fit the data. On the contrary, we have derived this theory through a very simple logical thought process, built-up from fundamental principles. We had no freedom to "fiddle" this theory. We discovered this theory – not invented it.

When I set out to construct this theory, I was only interested in seeing if a theory of gravity could be produced from the principle of the zero-energy universe. I had no idea that this "repulsive gravity" idea would pop-out, or that it would agree so closely with the observed acceleration of the expansion of the universe. This was a complete surprise.

Finally, I would like to end this chapter by recounting a funny true conversation I recently had with a friend of mine (who shall remain nameless!). I was explaining the principle of repulsive gravity, and I wondered what would happen to humanity if the force of gravity suddenly reversed so that all the objects on the face of the Earth flew upwards. I imagined us being trapped on the ceiling of our room – like in the *Poseidon Adventure*. My friend, though, had an ingenious solution to prevent himself flying upwards under reversed gravity: "There's no problem", he insisted, "You would just have to weigh yourself down with heavy weights!"

I hope you can see the flaw in my friend's logic.

8

HOW TO CREATE YOUR OWN UNIVERSE

So it appears that a modified theory of gravity could produce the accelerated expansion of the universe, and eventually result in a smooth, flat universe. But, no matter how convincing the argument, it is always best to see a theory working in practice. So, with this in mind, we are going to perform an experiment. Well, when I say **we** are going to perform an experiment, I actually mean that **you** are going to perform an experiment.

I have written a computer simulation of a universe which should run in a web browser on your computer. The simulation is based on the hypothesis which has been presented in this book, and will show the evolution of the universe from Big Bang to final radius.

You will need Java installed in your browser to order to run the simulation. If you do not have Java then you will need to go to www.java.com to get it installed.

The address you will need to enter in your web browser to go to the simulation is:

www.whatisreality.co.uk/universe

HIDDEN IN PLAIN SIGHT 2

Keep your eyes peeled because you may get various security pop-ups asking if you want to allow the simulation to run. Say yes.

Eventually, you should see a pop-up which asks for the strength of gravity in the simulation. Enter a value of 10:

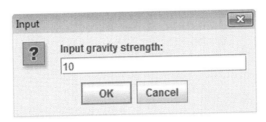

You will then see an explosive Big Bang in your newly-created universe, starting from just a point:

HOW TO CREATE YOUR OWN UNIVERSE

The universe will rapidly expand, before settling-down at its final equilibrium radius, its Schwarzschild radius:

You will notice a period of oscillation before the universe reaches this final radius. This is due to both the attractive and repulsive natures of gravity, acting like an oscillating "gravity spring".

It should be noted that I did not include the presence of an explosive force at the Big Bang. There is merely the requirement that the universe should expand to its Schwarzschild radius: gravity does it all.

You will notice that – even though the force of gravity is accurately modelled in the simulation – the universe does not collapse back to a point: it remains at a steady radius. The universe also appears smooth, with the mass being evenly-distributed. So the hypothesis presented in this book certainly seems like a simple and elegant way to create a universe.

You can click on the universe image at any point to create a new universe repeatedly. There is also a link at the bottom of the web page to download the actual code (if you are a Java programmer).

Have fun creating universes!

Is the universe a black hole?

Let's now turn our attention back to black holes.

In our analysis of black holes, we calculated that any object would become a black hole if all its mass was compressed to a smaller size than the Schwarzschild radius for that mass. At that point, the pull of gravity would become a one-way street and a singularity would inevitable occur. Well, that throws up an interesting scenario because the Schwarzschild radius of our universe appears to be larger than the radius of the universe. In other words, our universe fits the criteria for a black hole!

You might imagine that the density of the universe would not be enough to form a black hole. However, while it is true that small regions of space would require tremendous densities to form a black hole (Earth, for example, would have to be compressed to a 9mm radius), very large regions of space would require a far smaller density. For example, the average density of the region inside the event horizon of a supermassive black hole can be less than the density of water! This is because the Schwarzschild radius is proportional to mass, which is proportional to the cube of the radius. So as a region gets larger, its Schwarzschild radius increases at a much faster rate than its radius does. In fact, if you have a very large volume of mass at constant density, then it is **guaranteed** to become a black hole if the volume is large enough – **no matter how small the density!**[12]

This means that if we had a large volume of soft cotton wool, with a uniformly low density, it would collapse to form a black hole as long as the volume was large enough. Imagine a cotton wool black hole!

So people who say that the universe is arbitrarily large – or infinitely large – are surely saying that the universe has the characteristics of a region which will form a black hole. And certainly, if we consider the implications of inflation – which predicts a size for the entire universe which is a hundred sextillion times the size of the visible universe – then it would appear that such a universe would surely represent a black hole. This is what the equation tells us. To quote Steven Weinberg: "Our mistake is not that we take our theories too seriously, but that we do not take them seriously enough." That certainly seems to be the case here.

In this book, we have seen that a universe which reaches the size of its Schwarzschild radius is spatially flat. This would appear to indicate that a universe which is entirely contained within its Schwarzschild radius would be a closed universe. It is known that light cannot escape from a closed universe – just like a black hole. So there are clear parallels between a closed universe and a black hole, both in terms of their size and their behaviour.

When physicists consider this question of whether the universe is a black hole (which is not often), there appears to

---

[12] As Andrew Steane says in his book *Relativity Made Relatively Easy*: "At any given density we can attain the condition where the radius is less than the Schwarzschild radius just by making the radius big enough."

be a lot of skirting around the issue. Even though this appears to be a vitally important question, I get the impression that they feel uneasy about the possibility. I have seen criticisms that the universe cannot be a black hole because, of course, a black hole is defined to be a region inside a larger space, and there cannot be any larger space outside the universe. Though, of course, if you were an inhabitant of a black hole it would appear to you that your black hole was your entire universe, and there was no space outside. So I feel this is a very weak criticism.

But usually the conclusion is that the behaviour of matter inside the universe simply does not match the behaviour expected inside the event horizon of a black hole. Specifically, inside a black hole we would expect to find matter converging to a central singularity. However, in the universe, we find matter emerging **from** a central point.

So physicists make a comparison between our universe and a peculiar theoretical object called a *white hole*.[13] A white hole is like a time-reversed black hole. And, when you think about it, this makes a lot of sense. At the time of the Big Bang, the universe is believed to have been a point singularity, and from that time onwards matter was expelled from that singularity. In a black hole, matter is consistently attracted **to** the singularity — just the opposite of the Big Bang! Hence, the universe behaves as if it is a white hole, as if it was a black hole in reverse.

So the hypothesis of the universe being a black hole is usually rejected because, as I say, the behaviour of matter inside the universe simply does not match the behaviour

---

[13] Philip Gibbs, *Is the Big Bang a black hole?*
http://math.ucr.edu/home/baez/physics/index.html

expected inside the event horizon of a black hole. However – and this is a key point – **nobody knows what happens inside a black hole**. Nobody can ever know. It is impossible to get information out of a black hole, so no one knows what goes on inside a black hole.

But, as we have seen, the equations still point to the universe being a black hole. With this in mind, it would appear that when we look out on the universe, we are looking at the behaviour of matter inside a black hole. And rather than seeing matter collapsing towards a central singularity, we see space expanding outwards towards the Schwarzschild radius. I feel this is clear evidence that the hypothesis described in this book is correct. If the data contradicts our model of gravity, we should not ignore the data – we should modify the model.

With this in mind, on the basis of the hypothesis that was proposed in the previous chapter, we can see that the reason matter is propelled from a point singularity in our universe is because of the predicted repulsive behaviour of gravity inside a black hole. We do not need to consider our universe as a time-reversed black hole, or "white hole" – it is simply our incorrect model of the behaviour of gravity inside a black hole which is reversed, not time itself being reversed.

Enter the black sphere

This modified behaviour of gravity has huge implications for all black holes. Outside the event horizon, the behaviour of a black hole would appear completely unchanged: mass would continue to be sucked to the event horizon, and the black hole would eventually evaporate on schedule. There would continue to be an extraordinarily high amount of total mass inside the event horizon of the black hole – together with an accompanying extraordinary gravitational pull. However, the behaviour of that mass inside the event horizon would be completely different. The hypothesis of this book predicts repulsive gravity inside the event horizon of a black hole. Unfortunately, of course, we can never see inside the event horizon of a black hole.

However, we can simulate it.

In order to examine the behaviour of gravity inside a black hole, we are going to return to our universe simulation. So get your web browser going again and return to the computer simulation website. This time, though, we are going to enter a different value for the force of gravity. As black holes are so much smaller than the universe, distances between masses would be smaller and so the force of gravity would have a proportionately larger effect. So we are going to enter a value of 50 for the force of gravity:

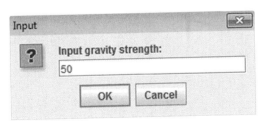

This has a truly remarkable effect. It appears that the effect of the increased strength of gravity is to concentrate the mass, and make it less spread-out. We now find that in our simulated black hole, all the mass now becomes concentrated at the Schwarzschild radius which, for a black hole, represents the event horizon!

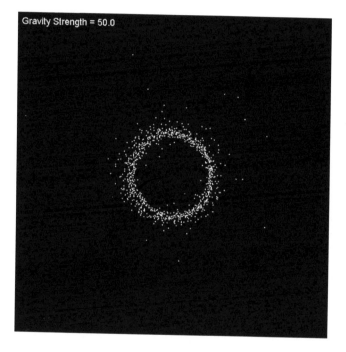

We can see the result would be less like a black **hole** and more like a black **sphere**!

For distant masses, attractive gravity is pulling those masses to the event horizon. And inside the event horizon, repulsive gravity is repelling interior masses to the event horizon. Basically, gravity moves all mass to the Schwarzschild radius. Hence, we observe a sphere of mass at the event horizon.

Note that the curvature of spacetime outside the event horizon is unaffected. To the exterior observer, the black hole interior would appear to be the same fearsome entity as before.

It has been suggested that this black hole structure has superficial similarities with a tornado. A tornado consists of a rapidly spinning hollow cylinder of debris. However, inside the cylinder is a comparatively calm and empty area (the "eye of the storm"). Similarly, a black hole is surrounded by a rapidly spinning disk of debris (remember the description of the accretion disk in Chapter Five). Also, the analysis which has been presented in this book suggests a calm and empty area inside the black hole — just like a tornado!

There are many remarkable features about this result. For a start, it appears to eliminate the singularity at the heart of the black hole. As we discussed in Chapter Five, if your physical theory has a singularity, it is a sign that your theory is wrong. Physics breaks down at a singularity. A theory with a singularity needs modification. Our current theory of black holes, derived from conventional general relativity, predicts the existence of singularities. However, the modified version of general relativity described in this book has the happy property of eliminating the singularity. The repulsive effect of gravity would fight against the tendency to collapse to a singularity.

Infinities do not occur in Nature. We have been too willing to accept the theory of singularities at the heart of black holes. The theory presented here has the unexpectedly pleasing side-effect of revealing why singularities — and associated infinities — would not occur inside black holes.

We can obtain another remarkable insight from this simulation. If you remember back to our discussion of the black hole information loss paradox in Chapter Five, you will recall the "firewalls" theory which suggests any infalling object has to be obliterated the moment it crosses the event horizon. This is in contrast with the "no drama" orthodox theory which predicts that the infalling observer will detect nothing out of the ordinary when he crosses the event horizon. According to AMPS, though (who proposed the firewalls theory), at the event horizon we will find this dramatic incineration process. But why should this dramatic region spring out of nowhere at the event horizon? A lot of physicists are uneasy about the firewalls proposal for that particular reason. However, as we can see, the modified theory presented in this book suggests that the mass of a black hole would all be concentrated at the event horizon. Any incoming object would be obliterated on the event horizon when it hits the mass of the black hole, solving the information loss paradox. The information of the object would be plastered around the mass at the event horizon, and emitted as Hawking radiation.

This aspect of the theory has similarities with the theory of *fuzzballs* based on string theory. According to the fuzzballs theory, there is no singularity inside a black hole. Instead, there is just a mass of strings, extending out to the event horizon. This is presented as a proposed solution to the information loss paradox as "the quantum information is not trapped at the centre but stays on the fuzzy surface and is emitted as Hawking radiation".[14] This solves the information paradox in much the same way as the theory proposed in this book.

Most crucially, **this also explains why the entropy of a black hole – the trapped information – is proportional to the surface area of its event horizon**. This is due to all the

---

[14] Wikipedia article on fuzzballs:
http://en.wikipedia.org/wiki/Fuzzball_(string_theory)

mass (and, therefore, its information) being evenly spread over the event horizon.

Clearly, this modified theory of gravity is an elegant solution to many of the most puzzling problems in physics and cosmology.

9

CONCLUSION

Let us travel back in time to 1859.

The French mathematician Urbain Le Verrier has examined the orbit of Mercury and discovered peculiarities: the orbit did not appear to agree with Newton's law of gravity. In order to explain this discrepancy, Le Verrier suggested that there was another small planet inside the orbit of Mercury which was dragging it off course. Le Verrier called this hypothetical planet "Vulcan". So, instead of suggesting that the law of gravity required modification, Le Verrier suggested that there was a physical cause behind the observation. However, despite the claims of several astronomers, the search for planet Vulcan drew a blank.

The solution to the peculiarity in the orbit of Mercury was only found when Einstein produced the theory of general relativity in 1915. So instead of there being a physical cause to the discrepancy, it was our theory of gravity which required modification.

Fast forward to the present day, and we find a very similar situation. Observations of the universe have revealed a discrepancy from the predicted behaviour: the expansion of the universe is accelerating. Physicists – like Le Verrier

150 years earlier – are once again loath to modify the theory of gravity. Instead, they are suggesting that there is a physical cause to this discrepancy: the existence of "dark energy". However, might it not be the case that the solution to the observed discrepancy could, once again, require a modification to the theory of gravity? This time, though, it would be general relativity requiring modification – not the Newtonian theory.

Maybe "dark energy" will turn out to be the new planet Vulcan?

In our discussion of the black hole information loss paradox, in was stated that either quantum mechanics or relativity must be wrong. There's no way round it. One of the two great theories of the 20th century **must** require modification. So which one is it going to be? Well, after reading this book you now know which side of the fence I lie on. The suggestion is that general relativity requires a fairly subtle and ingenious modification. However, it is never going to be an easy task proposing a modification to one of the great theories. At first glance, it might appear that the challenge is to "prove Einstein wrong". However, this is not the case. Einstein did not prove Newton wrong. Both Einstein **and** Newton were correct. Einstein simply built on the theory of Newton to produce a more sophisticated, more accurate theory. Einstein's result had to agree with Newton's result in simple cases.

And this necessity to agree with previous theories is a challenge (a stumbling block?) for all theories of modified gravity: they have to agree with Einstein's proven theory. Theories of modified gravity which attempt to explain the accelerated expansion of the universe have all tried to modify gravity at large scales. The problem is, as you then move back to consider smaller scales you find your modified measurements deviating from those predicted by Einstein's theory. And Einstein's theory always gets it right. It is very

CONCLUSION

hard to modify gravity at large scales for this reason. Einstein did such a good job.

The hypothesis of modified gravity presented in this book is really very ingenious because it avoids this drawback. Gravity is predicted to be modified only for objects whose entire mass is contained within their Schwarzschild radius. This eliminates almost every object contained within the universe! Ants, trees, planets, galaxies, galactic clusters, are all predicted to exhibit gravitational attraction precisely according to Einstein's theory. At long range, for every object, the hypothesis of modified gravity presented in this book is identical to Einstein's theory (an obvious exception is black holes, but more of that later).

However, the hypothesis predicts a modification to gravity at short range. Not all objects are affected: only objects whose entire mass is contained within their Schwarzschild radius are predicted to exhibit this repulsive gravity. Most importantly, the entire universe happens to be one of the very few objects affected. Hence, the acceleration of the expansion of the universe is predicted.

As a bonus, the hypothesis presents simple and elegant solutions to some of the greatest mysteries and paradoxes in cosmology. The universe is predicted to be necessarily flat – without inflation. Black holes (the only other objects whose mass fits entirely within their Schwarzschild radius) are no longer predicted to contain singularities, and a simple explanation is provided for black hole entropy being proportional to the area of the event horizon. The information loss paradox is also simply explained.

So this gives us a new way of looking at gravity. Gravity seems to play a much more central role in the universe than is realised. If you asked the average person what gravity does, they would probably say it pulls apples off trees. If they were pushed a bit further, they might say it is the force responsible for holding the planets in orbit. And this is clearly the case. As an attractive force capable of spanning the width of galaxies and beyond, it is responsible for holding the universe together.

However, according to this hypothesis, gravity does not just "pull like mad" as if it is competing in a tug-of-war competition. Instead, we find gravity operating as more of a "positioning" force. Its behaviour is more subtle than we give it credit. Objects are positioned quite precisely at their Schwarzschild radius: it is all done with the precision of a surgeon. We might interpret this as a coarse attractive force, which is how it reveals itself to us for most of the time, but really gravity is so much more.

If the hypothesis described in this book is correct then one equation – for the Schwarzschild radius – controls the positioning of **every** object in the universe under gravity. It gives the universe its overall structure, from being the motivation behind the Big Bang, to powering the accelerating expansion of the universe and specifying its eventual radius. But it also holds the planets in orbit around the Sun, it forms galaxies, it controls nuclear fusion in stars, it describes the behaviour of black holes, and it pulls apples off trees. One force. One equation.

It is truly the equation of the universe.

FURTHER READING

The Black Hole War by Leonard Susskind
Very readable. How Leonard Susskind was delighted to prove Stephen Hawking wrong about the black hole information loss paradox – only for the recent firewalls theory to apparently prove that Susskind was wrong.

Black Holes and Time Warps by Kip Thorne
Everything you ever wanted to know about black holes can be found in this impressive book.

The Inflationary Universe by Alan H. Guth
Everything about inflation, by the man who invented it.

http://www.spacetelescope.org/
Mind-blowingly beautiful images from the Hubble Space Telescope.

Edge of the Universe by Paul Halpern
The latest exciting developments in cosmology, including dark matter, dark energy, and the unusual structures which have recently been discovered in the universe.

Cosmology: A Very Short Introduction by Peter Coles
Good, solid, coverage of the basics.

Relativity, Gravitation and Cosmology by Ta-Pei Cheng
An intermediate book which is more mathematical, but not too heavy for the reader who wants to dig deeper.

ACKNOWLEDGEMENTS

Many thanks to Gregory Gabadadze, Massimo Porrati, Wayne Hu, Sean Carroll, and Raphael Bousso for their patient responses to my questions.

Thanks to Mark Gilder for the front cover photograph.

Thanks to Paul Noonan for the back cover photograph.

PICTURE CREDITS

Milky Way image is a public domain image courtesy of Wikimedia Commons.

CMB and shape of the universe image is courtesy of NASA/WMAP.

Planck image of the CMB is courtesy of the European Space Agency.

NAUTILUS-X, NuSTAR, and black hole image courtesy of NASA.

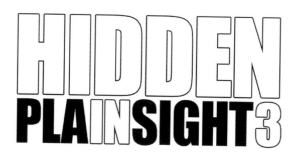

The secret of time

COMING SOON

Made in the USA
San Bernardino, CA
23 October 2013